AF576143

Size Effects on the Plastic Deformation of the BCC-Metals Ta and Fe

Zur Erlangung des akademischen Grades
Doktor der Ingenieurwissenschaften
der Fakultät für Maschinenbau
Karlsruher Institut für Technologie (KIT)

genehmigte
Dissertation
von

DIPL.-PHYS. DANIEL KAUFMANN
aus Bruchsal

Tag der mündlichen Prüfung: 23. November 2010
Hauptreferent: Prof. Dr. rer. nat. Oliver Kraft
Korreferent: Prof. Dr. Cynthia A. Volkert

Bibliografische Information der Deutschen Nationalbibliothek
Die Deutsche Nationalbibliothek verzeichnet diese Publikation in der Deutschen Nationalbibliografie; detaillierte bibliografische Daten sind im Internet über http://dnb.d-nb.de abrufbar.

1. Aufl. - Göttingen: Cuvillier, 2011
Zugl.: Karlsruhe, Univ., Diss., 2010

978-3-86955-654-3

Nonnenstieg 8, 37075 Göttingen
Telefon: 0551-54724-0
Telefax: 0551-54724-21
www.cuvillier.de

1. Auflage, 2011
Gedruckt auf säurefreiem Papier

978-3-86955-654-3

Acknowldegements

This work was created during my time as a PhD student at the Institute for Materials Research (IMF II) of the Karlsruhe Institute of Technology (KIT). At this point I thank all the people who helped me to finish my promotion successfully.

I thank Prof. Dr. rer. nat. Oliver Kraft for giving me the opportunity to do my PhD at the IMF II and for being the first examiner of my work.

I also thank Prof. Dr. Cynthia A. Volkert for being my supervisor at the beginning of my work and for being my second examiner.

Special thanks go to Dr. rer. nat. Reiner Mönig who became my supervisor after Cynthia Volkert went to Göttingen. His ideas and our discussions were important contributions to this work.

Thanks to my parents and my brother Andreas for the support they gave me.

Thanks to all the colleagues of the IMF II for the good working atmosphere, the fruitful discussions and the off the job activities.

Karlsruhe, November 2010 Daniel Kaufmann

Contents

List of Figures

Frequently Used Symbols and Abbreviations

fcc	face centred cubic
bcc	body centred cubic
hcp	hexagonal close-packed
SEM	Scanning electron microscope
EBSD	Electron backscatter diffraction
FIB	Focused ion beam
SS	short screw
LS	long screw
σ	normal stress
τ	shear stress
μ	shear modulus
$\vec{b}$	Burgers vector
b	norm of Burgers vector
ν	Poisson's ratio
E	Young's modulus
CRSS	critical resolved shear stress
T_m	melting temperature
T_{ath}	athermal temperature

Zusammenfassung

Größeneffekte können die Festigkeit eines Metalls beträchtlich erhöhen. Dieser Effekt wurde bislang für kubischflächenzentrierte (kfz.) Metalle in einer Vielzahl von Studien untersucht, jedoch ist wenig über solche Effekte in kubischraumzentrierten (krz.) Metallen bekannt. Krz. Metalle kommen in vielen technischen Anwendungen zum Einsatz, daher ist es wichtig, ihr größenspezifisches Verhalten zu kennen. Darüber hinaus sind mechanische Größeneffekte auch von wissenschaftlichem Interesse, da sich die zugrundeliegenden Mechanismen von denen in den kfz. Metallen grundlegend unterscheiden. Im Gegensatz zu kfz. Metallen, in denen Schrauben- und Stufenversetzungen annähernd gleich beweglich sind und die Beweglichkeit der Versetzungen nur schwach oder überhaupt nicht von der Temperatur abhängt, ist die Beweglichkeit von Schraubenversetzungen in krz. Metallen stark temperaturabhängig. Schraubenversetzungen bewegen sich in krz. Metallen über die thermisch aktivierte Bildung und Bewegung von Kinkenpaaren. Unterhalb einer materialspezifischen kritischen Temperatur (in der Literatur oftmals auch als athermische Temperatur bezeichnet) ist die Beweglichkeit der Schraubenversetzungen erheblich niedriger als die der Stufenversetzungen. In dieser Arbeit wurde der Mikrodruckversuch auf Ta und Fe angewandt, um das Verformungsverhalten dieser Metalle in kleinen Dimensionen zu untersuchen. Hierfür wurden Säulen, mit Dimensionen im Mikrometerbereich, mit einem fokussierten Ionenstrahl-Mikroskop (FIB) hergestellt und anschließend mit einer abgeflachten Diamantspitze in einem Nanoindenter auf Druck belastet. Die Experimente zeigen, daß krz. Metalle eine größenabhängige Festigkeit aufweisen. Der mechanische Größeneffekt, der in α-Fe beobachtet wurde, zeigt Ähnlichkeiten mit den Beobachtungen, die bei kfz. Metallen gemacht wurden und ist ausgeprägter als in Ta. Der Vergleich von Ta und Fe mit verschiedenen kfz. und krz. Metallen deutet darauf hin, daß die mechanischen Größeneffekte in krz. Metallen eine Temperaturabhängigkeit aufweisen. Die Unterschiede im mechanischen Verhalten von kfz. und krz. Metallen können den unterschiedlichen Beweglichkeiten von Schraubenversetzungen zugeschrieben werden. Die Ergebnisse der Mikrodruckversuche weisen darauf hin, daß Schraubenversetzungen den Verformungsprozess in den krz. Metallen kontrollieren. Weiterhin lässt sich aus den Experimenten schließen, daß Oberflächen die Beweglichkeit von Schraubenversetzungen in krz. Metallen erhöhen können. Das mechanische Verhalten kleiner Proben von krz. Metallen kann durch eine Summe aus dem größenabhängigen Verhalten von kfz. Metallen und einem temperaturabhängigen Term, welcher von der beschränkten Beweglichkeit der Schraubenversetzugen herrührt, beschrieben werden.

Abstract

Size effects can cause a significant increase in the strength of a metal. In the past, this effect has been extensively investigated for fcc metals but until today only limited knowledge exists on the size dependent behaviour of bcc metals. Bcc metals are used in many applications and, therefore, knowledge of their size scaling is of technological importance. Size effects in bcc metals are also interesting from a fundamental scientific point of view, since underlying dislocation processes of bcc metals differ from those of fcc metals. In contrast to fcc metals, where screw and edge dislocations have similar mobilities with little or no temperature dependence, the mobility of screw dislocations in bcc metals is strongly temperature dependent. In bcc metals, screw dislocations move by thermally activated kink pair nucleation and motion. Below the material dependent athermal temperature, their mobility is lower than that of edge dislocations. In this work, the deformation of Ta and α-Fe micropillars was studied using microcompression experiments. Pillars were produced using a focused ion beam microscope and were compressed in a nanoindenter using a flattened tip. The experiments show that there is a distinct size effect in bcc metals. The size effect of α-Fe is similar to the one found in typical fcc metals, and it is more pronounced than the one observed in Ta. A combination of this data with data obtained from fcc metals and from several other bcc metals indicates that the size dependence of bcc metals is temperature dependent. The differences in the mechanical behaviour between fcc and bcc metals can be attributed to the difference in mobility of screw dislocations in fcc and bcc metals. Results from microcompression experiments also indicate that screw dislocations control the deformation process. The results of these experiments can be explained by surface enhanced mobilities of screw dislocations. Overall, the mechanical behaviour of small bcc metals may be described by scaling laws as known from fcc metals and an additional temperature dependent component arising from the limited mobility of screw dislocations.

Abstract

1

Introduction

In this work, mechanical size effects of the bcc metals α-Fe and Ta have been investigated. One of the major applications of Ta is its usage as anode material in small electrolytic capacitors. These capacitors are used in microelectronic devices such as cell phones or in the electronic circuits of vehicles. Furthermore, Ta is an appropriate material for medical devices and prostheses since it does not react with human tissue and human bodily fluids. Ta is chemically and thermally resistant which makes it an appropriate material for thermal barrier coatings. Jet engines of aircrafts made of Ni-super alloys contain Ta to give them more thermal stability. Fe is a metal of great technological importance since it is the basis of steel and due to its ferromagnetic properties, Fe is used in generators, transformers, relays and electric motors. The capabilities of these two metals are manifold and knowledge of their mechanical properties is necessary to optimize their applications. Both metals can be utilized in microelectronic or in micromechanical devices. For instance, a mill that removes deposits inside a blood vessel would need a tool that does not react with human tissue and a very small electric motor. Applications like these make it necessary to investigate the mechanical behaviour of materials like Ta and Fe on the micrometer scale.

Furthermore, the mechanical behaviour of small-scaled samples is of scientific interest since it might give new insights into the deformation mechanisms and related size effects. From studies of many fcc metals it is known that mechanical properties of metals change when at least one dimension of the sample is reduced into the micrometer regime or below. A commonly observed phenomenon is the increase in strength with decreasing dimensions ('smaller is stronger') .

The resistance of a metal against plastic deformation can be described by the yield strength or alternatively by the flow stress which is usually measured at 0.2% plastic strain. Plastic deformation in metals occurs by the motion of one-dimensional lattice defects, the so-called dislocations. Is the sample large compared to the typical length scales of the dislocation networks, the dislocation density can be considered to be described by a homogeneous dislocation density. Thus, the yield strength of the metallic sample does not depend on its size. If one dimension of the sample is reduced into the micrometer regime, the yield strength becomes a function of sample size. In this regime, dislocations are not uniformly distributed anymore.

Although there are many studies in the literature that deal with mechanical size effects in fcc materials, the underlying mechanisms are still a matter of debate. This is even more pronounced for the size dependent behaviour of bcc metals.

The goal of this work is to investigate mechanical size effects of bcc metals and how they compare with results found for fcc metals. The comparison of both material groups will also lead to a deeper understanding of the involved dislocation mechanisms.

2

Background

2.1 Dislocations

Many properties of metals can be explained by considering their atomic bonds and crystal structure, for instance, their elastic constants, melting point, density etc.. Other properties can exhibit different values for the same metal. Examples are electrical conductivity, yield and fracture strength. The properties can be divided into two groups, the structure-insensitive and the structure-sensitive properties. The second group of properties depends on the defects that are present in the crystal structure of the metal. At temperatures above 0K, a crystal is not necessarily in thermodynamic equilibrium and will always contain defects. In this work, one-dimensional or line defects, the so-called dislocations, are of fundamental importance and will be discussed in more detail.

Dislocations are responsible for the phenomenon of slip by which most metals deform plastically. Two different types of dislocations can be distinguished, edge dislocations and screw dislocations. The two different types of dislocations are illustrated in figure (2.1). Figure (2.1(a)) shows the slip in a continuum that is produced by an edge dislocation. An important parameter of a dislocation in a crystal is its Burgers vector $\vec{b}$, which can be determined by the following procedure. Starting from an atom in a lattice a rectangle is drawn clockwise around the dislocation core whose edges contain the same number of atoms. The dislocation introduces a displacement into the crystal in a way that the rectangle will not be closed. The closure vector of the rectangle is the Burgers vector $\vec{b}$. For edge dislocations $\vec{b}$ is perpendicular to the dislocation line. In figure (2.1(b)) the slip produced by a screw dislocation is shown. The Burgers vector of a screw dislocation can be determined in an analogous way as the one of an edge dislocation. In this case the Burgers vector $\vec{b}$ is parallel to the dislocation line. If $\vec{b}$ is neither parallel nor perpendicular to the dislocations line, the dislocation is of mixed character and can be separated into an edge and a screw component.

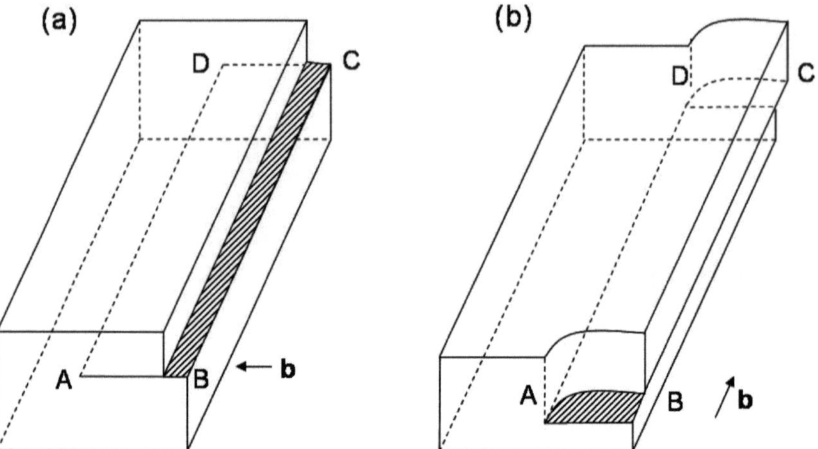

Figure 2.1: (a) Edge dislocation produced by slip in a continuum. Dislocation line is along AD, with $\vec{b}$ being perpendicular to AD. (b) Screw dislocation in a continuum. Dislocation line AD is parallel to $\vec{b}$. In both figures slip has occured over area ABCD.

2.2 Deformation of Single Crystals

In this work, the deformation behaviour of small single-crystalline metallic samples was investigated. The plasticity of single crystals is somewhat different from those of poly crystals, since the elastic and plastic properties of single-crystalline materials are usually not isotropic.

The extent of slip in a single crystal depends on the magnitude of the shearing stress produced by external loads, the geometry of the crystal structure, and the orientation of the active slip planes with respect to the shearing stresses. Slip occurs when the shearing stress on the slip plane in the slip direction reaches a threshold value, the so-called critical resolved shear stress (cf. figure (2.2)). The critical resolved shear stress can be determined according to

$$CRSS = \frac{P}{A} \underbrace{\cos\lambda \cdot \cos\phi}_{\text{Schmid factor}} \tag{2.1}$$

The Schmid factor is an important parameter when considering plastic deformation in single crystals. In fcc metals slip occurs on the slip system which exhibits the highest Schmid factor. In bcc metals slip does not necessarily occur on the slip system with the highest Schmid factor, since the Schmid factor is only a geometric factor, that does not take packing densities of slip planes or mobility of dislocations into account.

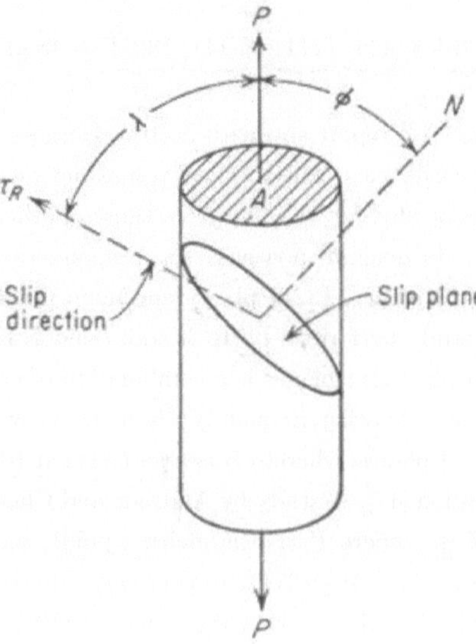

Figure 2.2: Diagram for the calculation of the critical resolved shear stress in a single crystal [1]. The load P acts on a single crystal with cross-sectional area A.

2.3 Plasticity in Bulk BCC Metals

BCC metals are of technological relevance and therefore the investigation of these materials is not only of scientific interest, but also valuable from a technological point of view.

Before investigating the small-scaled samples, it is useful to take a look at the deformation behaviour of bulk bcc metals because their plasticity differs significantly from the behaviour of fcc metals. A comparison of the glide systems in fcc and bcc metals demonstrates some of these differences. Glide of dislocations takes place on the closest packed crystallographic planes. The Burgers vector is usually the shortest connection between two atoms in the unit cell; i.e. $\frac{1}{2}[110]$ and $\frac{1}{2}[111]$ in fcc and bcc, respectively. In the case of fcc metals this leads to 12 glide systems of the following type.

$$\underbrace{\langle 110 \rangle}_{3(\text{symmetry of plane})} \cdot \underbrace{\{111\}}_{4(\text{number of planes})} \Rightarrow 12 \text{ glide systems}$$

For bcc metals the situation is more complex because there are three types of planes having almost equal packing density. The possible glide systems in bcc (cf. figure (2.3)) are

$$\underbrace{\langle 111\rangle}_{2}\underbrace{\{110\}}_{6}+\underbrace{\langle 111\rangle}_{1}\underbrace{\{211\}}_{12}+\underbrace{\langle 111\rangle}_{1}\underbrace{\{321\}}_{24}\Rightarrow 48 \text{ glide systems}$$

In addition, Reid et al. [14] report slip with a $\langle 100\rangle$-Burgers vector in niobium single crystals with a $\langle 111\rangle$ tensile axes. This atypical result may be due to elastic anisotropy. The anisotropy factors of Mo and Nb differ from those of other bcc metals, $A = \frac{2c_{44}}{c_{11}-c_{12}}$ are smaller than one. In general, however, the preponderant slip planes in bcc are the $\{110\}$ and the $\{211\}$ planes, $\{321\}$ planes and noncrystallographic slip planes are also found [15]. In a study performed by R. Maddin and N.K. Chen [16], it has been suggested that slip on the $\{321\}$ planes is a combination of cross slip processes on the $\{110\}$ and $\{211\}$ planes. Another frequently discussed view is that plasticity occurs only on $\{110\}$ and $\{211\}$ planes, whereby it is assumed that $\{321\}$ planes need a higher temperature for activation [17]. A study by Andrade and Chow [18] suggests that with an increasing ratio of $\frac{T}{T_m}$, where T_m is the melting point, the subsequently operative slip planes should be $\{211\}$,$\{110\}$,$\{321\}$, respectively. However, a comparison of the bcc metals in reference [15] indicates that the order is $\{110\}$,$\{211\}$,$\{321\}$. Noncrystallographic slip seems to occur only at higher temperatures or low strain rates and can be attributed to cross slip of screw dislocations. Slip in bcc metals obviously is quite complex and there is no consensus about the possible slip systems. For the identification of glide planes, later in this work (cf. chapter (5.4)), all suggested planes have been taken into account, namely $\{110\}$, $\{211\}$ and $\{321\}$. One fact, that makes slip in bcc metals even more complex, is that slip does not follow Schmid's law [2, 19, 20].

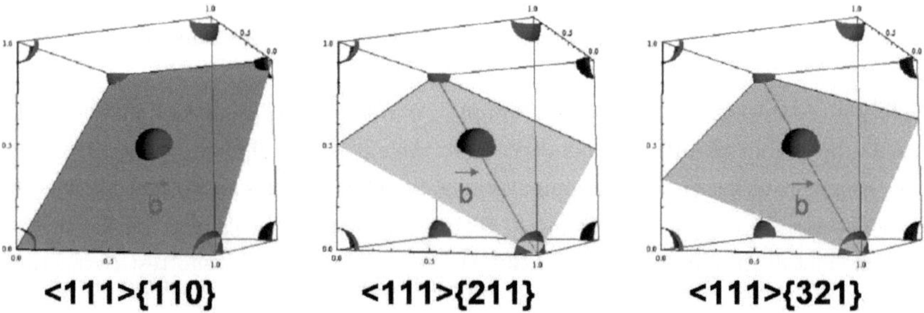

Figure 2.3: Glide systems in the unit cell of a bcc metal. The line represents the direction of the Burgers vector $\vec{b}$

2.3.1 Screw Dislocations in BCC Metals

Besides the differences in glide systems, there are differences in the structure of screw dislocations between fcc and bcc metals. In fcc metals, screw and edge dislocations move with approximately the same velocity when they are exposed to shear stresses. The structure of screw dislocations in bcc metals is very complex and their motion plays an important role in the plasticity of bcc metals. The mobility of the screw dislocations is severely restricted by the particular structure of their cores. Around a $\langle 111 \rangle$ direction, the cubic structure has a threefold symmetry. This direction contains three $\{110\}$ and three $\{112\}$ planes which are potential dissociation and slip planes in the bcc structure.

The core structure of a $\frac{1}{2}[111]$ screw dislocation depends on the specific bcc metal as has been shown by K. Ito et al. [21]. The screw dislocation core of Mo differs fundamentally from the one of Ta. In Mo, the core spreads asymmetrically into the planes $(\bar{1}01)$, $(0\bar{1}1)$ and $(\bar{1}10)$ that belong to the [111] zone. This asymmetry leads to two energetically equivalent core structures, the core strucure therefore is twice degenerated. This core structure can be perceived as a generalized splitting into three fractional dislocations with screw components $\frac{1}{6}[111]$. In contrast, the Ta core is non-degenerated since it spreads symmetrically across the $(\bar{1}01)$, $(0\bar{1}1)$ and $(\bar{1}10)$ planes and can be regarded as splitting into six fractional dislocations with screw components $\frac{1}{12}[111]$.

Atomistic methods yield that the extension of the core structure is two to three Burgers vectors [2]. The non-planar spreading of the screw dislocation cores is the main reason for the breakdown of the Schmid law in bcc metals. The core has to be transformed in order to move the dislocation and this transformation may be affected by shear stresses in the slip direction acting on planes other than the slip plane, as well as by shear stresses in the direction perpendicular to the Burgers vector [19,22]. Not only the complex structure of the screw dislocations but also their ability to cross slip affects the plastic behaviour of bcc metals. Screw dislocations with $b = \frac{1}{2}\langle 111 \rangle$ can glide on three $\{110\}$ and three $\{211\}$ planes whereas in fcc metals a screw dislocation with $b = \frac{1}{2}\langle 110 \rangle$ can only glide on two planes of the $\{111\}$ type. Therefore, cross slip is observed more often in bcc than in fcc metals. In bcc metals, the motion of screw dislocations naturally involves more cross slip. The motion of a dislocation on a glide plane is not only determined by the resolved shear stress but also by the mobility of the dislocation [23, 24]. One example for this is the frequently observed "wavy-slip" which is periodic cross slip between two glide planes.

The origin of the specific plastic behaviour of bcc metals can be traced down to the aforementioned non-planar extended core configuration of the screw dislocations of these materials. This implies a large value for the lattice-friction Peierls stress, which has to be overcome by the gliding screw dislocations with the help of thermal activation [25]. The high Peierls stress is not due to the nature of the chemical bonding, as in diamond cubic semiconductors, but to a particular symmetry property of the bcc lattice. Non-screw segments do not share this particular property and their mobility can be described in much the same way as in fcc crystals (cf. [2]).

The motion of the screw dislocations in the bcc lattice occurs by the thermally assisted formation and migration of kink pairs. It is noteworthy, that the mobility of non-screw dislocations is high and weakly temperature-dependent. In contrast, the mobility of screws is comparatively very low but strongly temperature-dependent. The ratio of edge to screw velocities is, therefore, decreasing with temperature, going from orders of magnitude at very low temperatures to close to unity at a certain temperature such that the activation energy for overcoming the lattice friction is entirely provided by thermal fluctuations. This 'athermal temperature', T_{ath}, is about $0.15 \cdot T_m$, where T_m is the melting temperature. It is a measure of the total height of the Peierls barrier. As a consequence, the mechanical behaviour of bcc crystals becomes very similar to that of fcc crystals at temperatures above T_{ath} [2]. Different values for the athermal temperature T_{ath} can be found in literature [2,26], but values of $0.15-0.2 \cdot T_m$ are a good estimate. A more precise way of determining the athermal temperature is to measure the specific temperature where the yield strength of a bcc metal becomes independent of temperature.

As mentioned above, the screw dislocations move by thermally activated kink pair nucleation. Only at zero temperature, the screw dislocations line moves as a whole across the Peierls relief. The corresponding stress, τ_p, is the stress needed to transform the 3D core into a 2D glissile structure. At a finite temperature and under an effective stress $\tau^* < \tau_p$, the saddle-point configuration consists of a small bulge (cf. figure (2.4)), whose formation requires an activation enthalpy $\Delta H(\tau^*)$. When this critical configuration becomes unstable, it expands laterally. As the non-screw segments have a mobility comparable to that of dislocations in an fcc lattice, their sideway motion under the effective stress τ^* is very fast. The screw dislocation is then transferred to the next Peierls valley where it takes its sessile configuration of minimum energy. The consecutive steps of screw dislocation motion can be seen in figure (2.4). The screw dislocation is dissociated in a threefold manner according to Escaigs elastic model [27]. The velocity of a screw dislocation can be written as

$$v = v_0 e^{\frac{-\Delta H(\tau^*)}{kT}} \tag{2.2}$$

where the prefactor v_0 gives the velocity that would be obtained, if each attempt had a probability of success equal to unity, that means that v_0 represents the velocity of the dislocation above T_{ath}. The attempt frequency Γ can be approximated by the constant value $\nu_d b/l_c$ [2], where ν_d is the Debye frequency and l_c is the critical length of a kink pair. Γ is approximately the time taken by a sound wave to propagate over the critical length l_c. v_0 can now be approximated by the following relation [2, 28]

$$v_0 \approx b\Gamma\frac{L}{l_c} = \frac{L\nu_d b^2}{l_c^2} \tag{2.3}$$

The sideway kink motion is very fast at low temperatures due to high stresses and high kink mobility. The maximum number N of kink pairs along a dislocation line of the length L is $N = L/l_c$. L is the free-length of screw character between two strong obstacles, as for instance pinning points. Kink pair nucleation can occur on different glide planes. When the kinks move along the dislocation their interaction may serve as a dislocation source (cf. chapter (2.3.2)). From equations (2.2) and (2.3) can be seen that the screw dislocation velocity depends on temperature, flaw density and the Debye frequency (respectively the Debye temperature) of the material.

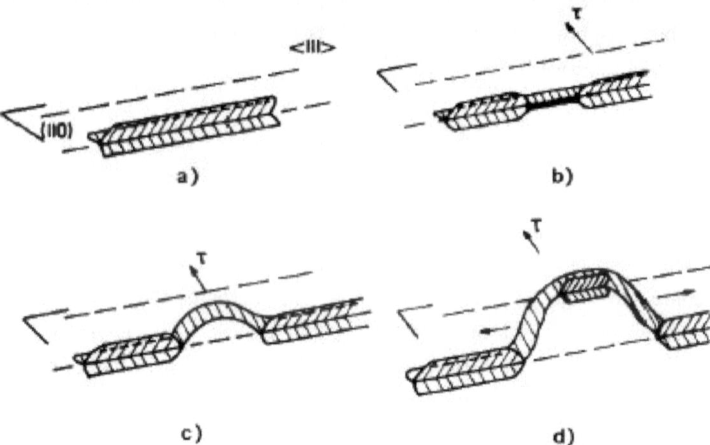

Figure 2.4: Successive steps of the formation and expansion of a kink pair on a screw dislocation in a bcc crystal within a 'dissociation-recombination' model [2]

2.3.2 Dislocation Sources in BCC Metals

Besides CRSS and the mobility of dislocations, the availability of dislocations affects plasticity in particular in the case of small specimens and it is therefore worthwile to look into possible dislocation multiplication processes in bcc metals. The generation and activation of dislocation sources determines deformation and hardening behaviour of a metal fundamentally. Several mechanisms can serve as dislocation source, the most known is probably the Frank-Read source. This multiplication process can be found in many dislocation text books (eg. [1, 15]) and it will therefore not be discussed here. Besides the Frank-Read source, other mechanisms provide dislocations for the deformation. Double cross-slip can be an effective dislocation source [3]. This mechanism was suggested by Koehler [29] and Orowan [30] and was experimentally first observed by Johnston and Gilman [31]. The double cross slip mechanism is sketched in figure (2.5). A screw dislocation (a) moves on its glide plane, which is identical with the image plane. Cross slip of a segment of length L results in two immobile superjogs J acting as pinning agents. The segment then multiplies similarly to the Frank-Read source. Simple line tension arguments show that both sources can act only, if the segment L is larger than a critical value

$$L_c = \frac{\mu b}{\tau} \tag{2.4}$$

where μ is the shear modulus, b the absolute value of the Burgers vector and τ is the local component of the acting stress. Characteristic values of L_c are in the range of 100-200nm. In stage (c) of the double cross slip mechanism, the branches adjoining the jogs have to pass each other on their parallel planes. This is only possible if the height of the jogs, i.e. the distance of the parallel glide planes, is larger than a critical value (dipole opening criterion)

$$h_c = \frac{\mu b}{8\pi(1-\nu)\tau} \tag{2.5}$$

with ν being the Poisson's ratio. In bulk material h_c is approximately 20 times smaller than L_c. Double cross slip as a dislocation source has also been observed in 3D dislocations dynamics simulations [32].

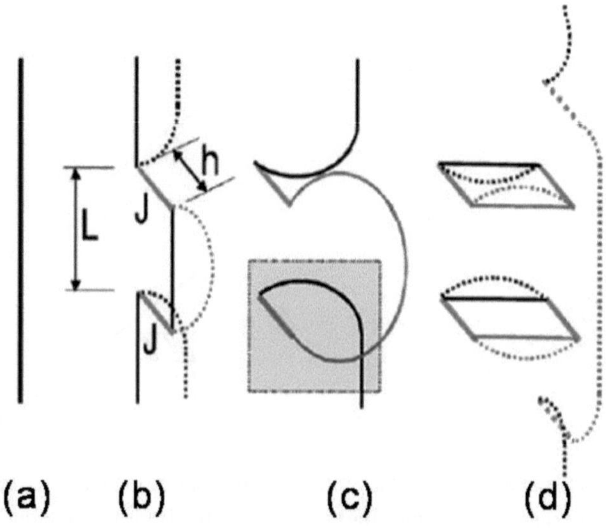

Figure 2.5: Double cross-slip as a mechanism for dislocation multiplication [3]

Double cross-slip is not the only mechanism that leads to an enhanced availability of dislocations in bcc metals. Another dislocation multiplication process in bcc metals can be attributed to the interaction of kinks on a screw dislocation line. Discrete dislocation dynamics simulations by W. Cai et al. [4] show that this interaction results in the formation of dislocation loops, so-called debris dislocations. The debris dislocation loop formation starts when two kink pairs form spontaneously on two different $\{110\}$ planes. As can be seen in figure (2.6(a)), when such kinks move towards each other and collide, they cannot recombine and create a pinning point. Because they are pushed towards each other by external stress, the kinks are now constrained to move together. In such cases, the forces on the two cross kinks act in the opposite directions along the line and the kinks can slow down or halt their coupled motion altogether. Such elementary kink-jog (or cross kinks) pairs grow in size when more kinks pile-up on either sides of the initial pinning point (A) forming superjogs, as shown in figure (2.6(b)). Due to ease of cross-slip, the kinks in the pile-ups on the two sides of the pinning point may belong to different planes. Consequently, projections of the two developing pile-ups on the $\langle 111 \rangle$ plane appear similar to two random walk trajectories originating from the same point (initial cross kink) on a two-dimensional lattice. If two such trajectories cross each other, the dislocation line may reconnect by recombination of kink pairs. As a result, the two pile-ups are now reduced in size leaving behind a prismatic (debris) loop L, as shown in figure (2.6(c)).

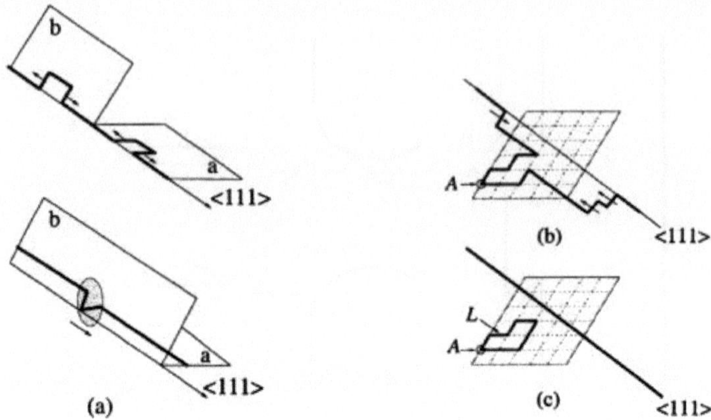

Figure 2.6: Schematic representation of the formation of (a) two kinks forming an elementary superjog, (b) more kinks joining the superjog, and (c) debris loop L formation with the primary dislocation break away from the self-pinning point A [4].

The special structure of screw dislocations gives rise to another dislocation multiplication mechanism which was identified by molecular dynamics simulations [5]. In the simulation a mixed dislocation with Burgers vector $b = 1/2\,[111]$ is introduced on the $(01\bar{1})$ glide plane. Under an applied shear stress the dislocation reorients itself and becomes a pure screw dislocation, making the subsequent dislocation behaviour insensitive to initial orientation. The edge segments leave the pillar comparatively fast and a pure screw is left behind. As the screw dislocation moves forward a cusp forms on the dislocation line. The cusp evolves to a dislocation loop. As the loop grows larger the two sides of the loop eventually leave the pillar, creating three dislocation lines in the pillar. Two of these dislocations have the same Burgers vector and move in the same direction as the initial dislocation, while the third dislocation has the opposite Burgers vector and therefore moves in the opposite direction. The same mechanism can now take place on the three dislocations now, so that a single nucleation event in a bcc pillar can trigger dislocation activities for a prolonged period. The cusp on the dislocation line forms because of two kinks on different glide planes interact with each other. This mechanisms bears resemblance of the one suggested by W. Cai et al. [4]. The evolution of this dislocation multiplication is illustrated in figure (2.7).

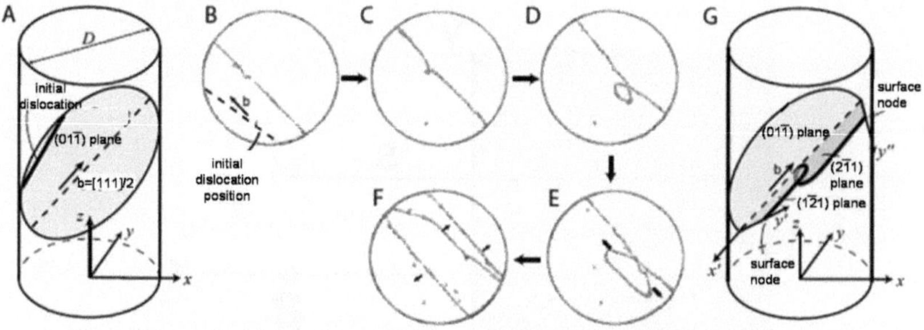

Figure 2.7: Dislocation multiplication by a cusp forming on a screw dislocation [5].

2.3.3 Influence of Image Stresses on Dislocations

Small-scaled samples have a very high surface to volume ratio which can give rise to effects that can be attributed to the surface. One important effect are image stresses that are active in the vicinity of the surface. These stresses interact with the stress field of a dislocation and therefore affect its motion and the deformation behaviour of the sample [33]. Figure (2.8) shows a right-handed screw dislocation which is located parallel to a free planar surface. At the surface, boundary conditions have to be satisfied, for instance that no forces can act normal to a free surface,

$$\sigma_{ij} n_j = 0 \tag{2.6}$$

with n_j the components of $\vec{n}$, the local surface normal. This boundary condition leads to $\sigma_{xz} = 0$ at the free surface and is satisfied if the self stress of the screw dislocation and that of an imaginary screw of the same strength and opposite sign are superimposed. The image screw dislocation is located at the mirror position outside the solid of the real screw dislocation. The image stress acting on the core of the dislocation can be determined by

$$\sigma_{yz} = \frac{\mu b}{4\pi s}. \tag{2.7}$$

This means that the screw is drawn towards the surface by a force F_x acting on a dislocation segment of the length L which can be calculated according to

$$\frac{F_x}{L} = \frac{\mu b^2}{4\pi s} \tag{2.8}$$

The image forces become more prominent as the sample dimensions are decreased, because the fraction of the sample, that is affected by the image forces, becomes larger. In bcc metals image stresses can generate kinks on the screw dislocations and thereby increase their mobility [34].

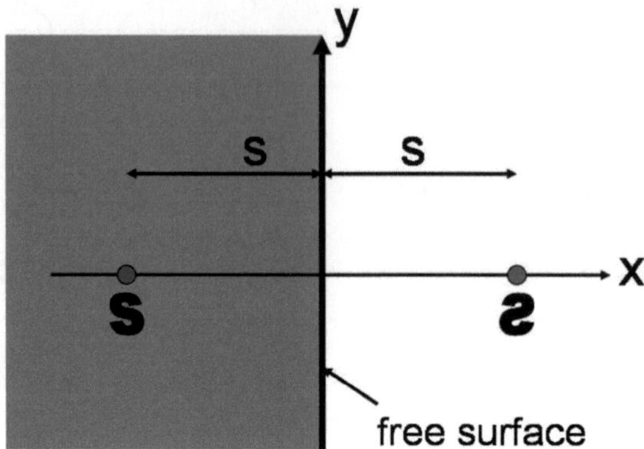

Figure 2.8: Screw dislocation in the vicinity of a free surface and its image screw.

2.4 Mechanical Size Effects

Plastic deformation of metals is primarily governed by dislocation mechanisms. The resistance against plastic deformation is low if the dislocations have a high mobility, do not strongly interact and dislocation sources can be easily activated. In a bulk sample, the number of dislocations is very high (in well grown single crystals $\approx 10^4 \mathrm{cm}^{-2}$, in severely deformed crystals up to $10^{12} \mathrm{cm}^{-2}$). Consequently, the mean spacing between two dislocations is in the range between $\approx$ 10nm and $\approx$ 100µm. In real samples dislocations typically interact and form networks with characteristic spacings on the micrometer range. If at least one dimension of the sample is reduced into the micrometer regime or below, the mechanical behaviour of the metal changes. In this regime the dislocations cannot be considered as uniformly distributed anymore, leading to a stochastic behaviour of plasticity in small dimensions. The decrease in dimensions is acompanied by an increase in strength, often referred to as 'smaller is stronger' [7,35–38]. The onset of the size dependent behaviour is on the order of several tens of micrometers. It is unlikely that there is a sharp transition from bulk to size dependent behaviour since the mechanical size effect is affected by crystal structure, initial dislocation density and surface to volume ratio of the sample.

In the past, the mechanical size effects were studied by a large number of experiments where different techniques have been applied. Some of these techniques will be discussed in more detail in this chapter. As a first example of a mechanical size effect, the Hall-Petch-Relation will be discussed where the size dependent behaviour of the material arises from the size of the grains in a poly-crystalline material. Furthermore, it will be shown that different dislocation mechanisms can lead to different scaling behaviours.

2.4.1 Hall-Petch-Relation

The Hall-Petch-Relation describes quite generally the grain-size dependence of the yield strength in many metallic materials [39].

$$\sigma_y = \sigma_0 + \frac{k}{\sqrt{gs}} \tag{2.9}$$

with σ_y =yield strength, σ_0 =yield strength of single crystalline material, k= positive constant and gs= grain size.
The yield strength of the material increases as the grain size decreases. Equation (2.9) would imply that the yield strength can be increased until infinity. In reality the Hall-Petch-Relation is only valid as long as the grain size is larger than a critical value. This critical value depends on the material and is usually in the range of several tens of nanometers. If the grain size is below the critical value, equation (2.9) is not valid anymore and the material may even soften (inverse Hall-Petch relation) [40]. The mechanical behaviour of nanocrystalline materials is subject of current research and many aspects are still not fully understood.
There are different approaches that can explain the empirical Hall-Petch-Relation, the pile-up of dislocations at the grain boundary will be discussed in the following.

Dislocation Pile-up Model

Grain boundaries act as obstacles for dislocation motion. Dislocations, generated by a Frank-Read source in the middle of a grain of diameter gs, will pile-up as they face the grain boundary. The applied shear stress is increased at the grain boundary due to the superposition of the stress-fields of the dislocations in the pile-up. The effective shear stress at the grain boundary is $n\tau_i$, where τ_i denotes the resolved shear stress. The number of dislocations in a pile-up can be determined by

$$n = \frac{\kappa\pi\tau_i gs}{4\mu b} \tag{2.10}$$

μ being the shear modulus, b the amount of the Burgers vector and $\kappa = 1$ for screw dislocations, $\kappa = 1 - \nu$ (ν is the Poisson's ratio) for edge dislocations. For a given shear stress more dislocations will pile up the larger the grain. When a critical stress τ_c is reached, dislocations are nucleated in the adjacent grain and macroscopic deformation can occur.

$$\tau_c = n\tau_i = \frac{\kappa\pi\tau_i^2 gs}{4\mu b} \tag{2.11}$$

The resolved shear stress which is necessary to overcome the influence of the grain boundary can be written as the difference of the applied shear stress and the shear stress which is necessary to move a dislocation through the lattice.

$$\tau_i = \tau - \tau_0 \tag{2.12}$$

(2.12)⇒(2.11)

$$\tau_c = n\tau_i = \frac{\kappa\pi(\tau - \tau_0)^2 gs}{4\mu b} \tag{2.13}$$

⇒

$$\tau = \tau_0 + \sqrt{\frac{4\tau_c \mu b}{\pi gs}} = \tau_0 + \frac{k'}{\sqrt{gs}} \tag{2.14}$$

2.4.2 Mechanical Size Effects in Thin Metal Films

Another example for size dependent strengthening is the plasticity of thin metal films where a reduction of one dimension leads to modified dislocation reactions.
The yield strength of a thin metal layer on a substrate depends on the thickness of the film. This behaviour is commonly investigated by wafer curvature measurements. For these measurements a thin metal film (from several hundred nanometers to several micrometers) is deposited onto a comparatively thick substrate. This compound is exposed to a temperature change and the film is strained because of the difference in thermal expansion coefficients. The mechanical stresses in the film can reach very high values (e.g. several hundred MPa for a 1µm Al-layer on Si [41]). Plastic deformation of the film requires motion and multiplication of dislocations. The interface of the substrate and the film is an insurmountable obstacle for the dislocations. This results in a special dislocation configuration. The dislocation can now be separated into two segments (cf. figure (2.9)), the mobile threading segment t and an immobile misfit segment m which is deposited at the interface [42]. For this kind of dislocation motion a critical shear stress is required as shown by Matthews et al. in 1970. Nix was able to quantify the biaxial stress σ in a thin film in 1988 [37],

$$\sigma = \frac{1.9b}{4\pi(1-\nu)h} \cdot \frac{2\mu_f\mu_s}{(\mu_f + \mu_s)} \ln\left(\frac{\beta h}{b}\right) \tag{2.15}$$

where b is the Burgers vector, ν is the Poisson's ratio, h is the film thickness, μ_s and μ_f are the shear moduli of the film and substrate material and β is a constant.

It is evident from equation (2.15) that the yield strength of the film scales roughly with h^{-1} independent of grain size. Wafer curvature measurements of small grained poly-crystalline films show stresses that are far above the stresses predicted by equation (2.15) [43]. It has been argued that the single dislocation mechanism plays a dominant role when film thicknesses below 100nm are considered. In films thicker than 100nm

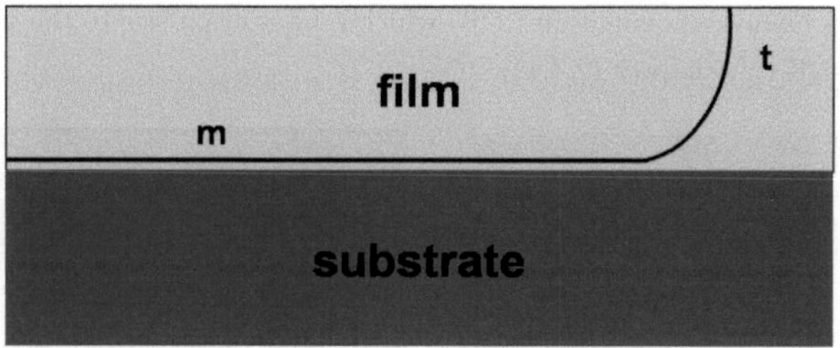

Figure 2.9: Dislocation motion in a thin film. A mobile threading segment t is moving along the film and deposits a misfit segment m at the interface film-substrate.

other processes have to be active. In equation (2.15) the influence of the grain size is not considered [44]. According to the Hall-Petch relation the grain size gs contributes to the yield strength proportionally to $\frac{1}{\sqrt{gs}}$. If the Hall-Petch-Relation is considered separately the stresses predicted for the wafer curvature experiments are smaller than those measured in the experiment [45]. This suggests that in thin films both mechanisms are active, the constraint from the film thickness and from the in-plane grain size.

2.4.3 The Microcompression Technique

In 2004, the microcompression technique was developed by Uchic et al. in order to investigate small metallic volumes [6]. Pillars with diameters in the range of several tens of micrometers down to $\approx$ 100nm are compressed in a nanoindenter equipped with a flattened tip. The pillars are usually machined with a Focused Ion Beam Microscope (FIB), but there are several studies where other techniques like etching [46] or nanoimprinting [47] were used. During compression, force and displacement are measured which can be converted into engineering stress and engineering strain by using the initial dimensions of the pillars. True stress and true strain cannot be easily determined due to the complex deformation behaviour of the pillars. Microcompression is a convenient method to quantify size effects, to compare different materials and to shed light on the involved dislocation mechanisms.

In ductile materials, the increase in strength of the pillars with decreasing diameter can often be phenomenologically described by an exponential relation between column diameter and strength,

$$\sigma_y(d) = \sigma_0 + c \cdot d^{-\beta} \tag{2.16}$$

with σ_y being the yield strength, σ_0 the bulk yield strength, c is a positive constant, d is the pillar diameter and β the size effect exponent. For $\beta = \frac{1}{2}$ equation (2.16) equals the Hall-Petch behaviour (equation (2.9)) which relates grain size to the yield strength of bulk materials (cf. chapter (2.4.1)).

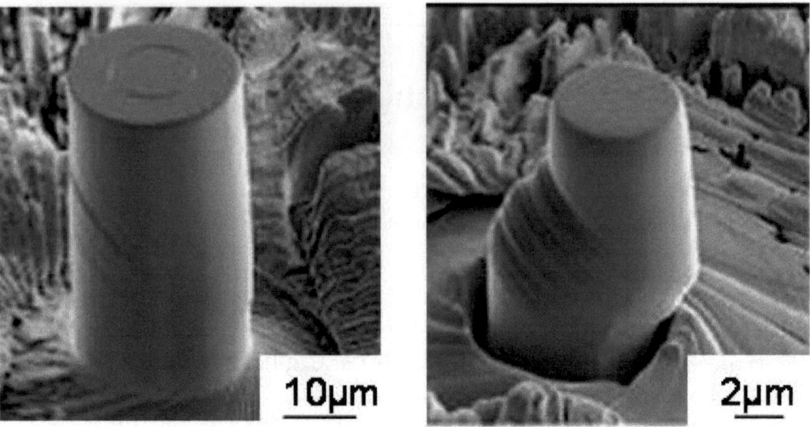

Figure 2.10: SEM micrographs: Pure Ni pillars after compression [6].

Microcompression testing has become a frequently used method to investigate the deformation behaviour of small-scaled samples. To date, more than 70 publications can be found in which mechanical size effects were investigated by microcompression testing. A brief overview over the relevant microcompression studies will be given in the following. The studies mainly focus on fcc metals due to the relatively simple deformation behaviour of these metals and their meaning in technical applications.

FCC Metals

The initial microcompression experiments focused on fcc metals because of their relatively simple lattice structure and the resulting simplicity of plasticity. FCC metals exhibit only 12 possible glide systems which are all of the $\langle 110 \rangle \{111\}$ -type. The activation of glide systems follows Schmid's law (equation (2.1)) which says that the glide system with the highest resolved shear stress will be activated. The first microcompression experiments were performed on Ni, which was oriented for single slip [6]. The phenomenon that has gained the most attention is the increase in flow stress with respect to the reduction of sample dimensions. The samples that have been investigated have diameters in the range from several hundred nanometers to a few tens of micrometers. This 'smaller is stronger'-behaviour has been demonstrated in Ni [6,48–50], Au [9,51–53], Al [38,54] and Cu [38,55,56]. In the framework of this work, tests have also been performed on Cu (chapter (5.1.1)). The increased flow stress results from

both increases in the proportional limit and tremendous strain hardening at micro to moderate strains at rates that are often greater than stage II hardening [49, 57]. With increasing strain the strain-hardening rate decreases rapidly. Due to the fact that strain hardening is size-scale dependent, smaller samples exhibit higher strain-hardening rates [9,49,50]. This leads to the observed high flow stresses of pure metallic crystals having sub-micrometer dimensions. For instance, it has been reported that 200nm $\langle 111 \rangle$ Ni pillars can sustain stresses in excess of 2GPa [48, 50], whereas pillars with diameters larger than 20µm show bulk behaviour with respect to strain hardening and flow stress. By reducing the sample size, the flow stress can increase by a factor of approximately 40 and the size-affected yield and strain-hardening behaviour becomes more stochastic in nature. In particular small samples often exhibit elastic or quasi-elastic load segments separated by so-called strain bursts which are intervals without any strain-hardening [6]. The strain bursts occur above a critical stress value and the stresses at which strain bursts can be observed as well as the magnitude of the strain bursts are distributed randomly. The relationship between diameter d and flow stress σ can be empirically described by a power law, according to equation (2.16).
This relationship fits well for the whole diameter range from several hundred nanometers to tens of micrometers. Power-law exponents, β, between 0.61 [9] and 0.97 [10] have been observed so far. These values are between the values observed for grain-size (n=0.5, cf. equation (2.9)) and thin film strengthening (n=1) (cf. equation(2.15)). Comparing the different studies is not straightforward since the microcompression experiment does not follow a standard procedure. Depending on the research group, the pillars are fabricated by lathe milling or top down milling (cf. chapters (3.2.1) and (3.2.2)). Furthermore, the diameter of the pillar can be measured at the top or at midheight of the pillar, the latter is done in order to get the mean stress which is active in the pillar. The flow stresses are taken at different strains, ranging from the proportional limit [54] to 10% plastic strain [52, 53].
Data obtained for different metals from these studies can be normalized by the shear modulus μ and plotted in a double logarithmic diagram with pillar diameter on the x-axis and normalized flow stress on the y-axis (figure (2.11)). This normalization procedure can be considered as normalization with respect to the theoretical shear strength σ_{theo}, which can be estimated according to

$$\sigma_{theo} = \frac{\mu b}{2\pi a} \tag{2.17}$$

with a the interplanar spacing and b the Burgers vector [58]. When applying this procedure to the data, the values follow a trend line with the slope being the power law exponent $\beta \approx 0.6$. The description of the data with a single exponent may be somewhat problematic since it is possible that depending on the size of the pillar, different strengthening mechanisms are active, for instance dislocation interaction in large samples and dislocation starvation in small samples. Furthermore, the initial dislocation density plays an important role for the plasticity of the sample as can be seen from the studies of Bei et al. [59] and the 3-D discrete dislocation dynamics study by Rao et al. [60]. The good agreement of the fcc studies might be due to the fact that all studies were performed on samples with comparable initial dislocation densities or that possible differences caused by different mechanisms are smeared out by plotting the data on a double logatihmic scale.

Multiple-slip orientations [50, 52, 53, 56] as well as single-slip orientations [6, 9, 49, 54] show size dependent strengthening. The relative importance of the crystal orientation vanishes as the sample dimensions are decreased into the micrometer regime, because stress strain curves of multiple-slip orientations and single-slip orientations become qualitatively similar when size-affected strain hardening dominates plastic flow.

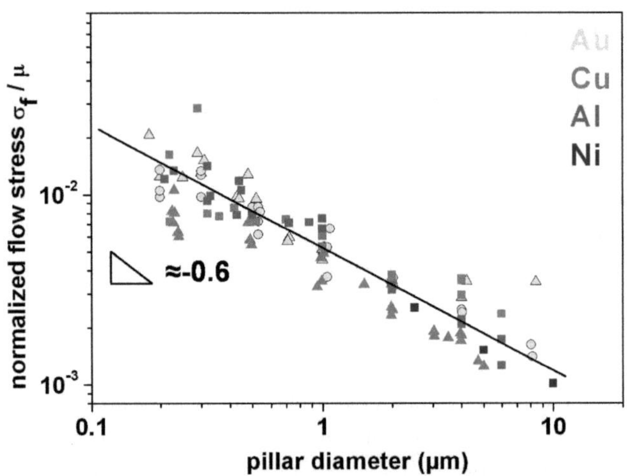

Figure 2.11: Microcompression data from several fcc metals. Flow stress is normalized by shear modulus μ [7]

Simulations of Small Scale Plastic Deformation of FCC Metals

The Hall-Petch Relation (cf. chapter (2.4.1)) and the example of the thin metallic films (cf. chapter (2.4.2)) show that the scaling behaviour in the strength of a metal can give information about the active dislocation mechanisms. Computational simulations of the deformation of small metallic samples help to identify the mechanisms that are active during deformation and that lead to a size dependence.
A powerful tool to investigate mechanical size effects theoretically is the Discrete Dislocations Dynamics (DDD) method. This technique describes the temporal evolution of a predetermined distribution of dislocations under the influence of predefined boundary conditions. From these investigations it is possible to see what happens inside a pillar when it is deformed. Dislocation processes like interaction, motion and source activation can be simulated and therefore help to understand the underlying mechanisms of the mechanical size effects [38, 61–68].
Senger et al. performed DDD-simulations for tensile tests on small scaled Aluminum samples [69]. The initial dislocation structure consisted of randomly distributed Frank-Read sources having a uniform length of 220nm. The initial dislocation density had a fixed value, so that the number of initial Frank-Read sources scaled with the volume of the investigated sample. Although the initial parameters are somewhat artificial, this study gives some new insights into the deformation behaviour of small scaled samples. In small pillars, single dislocations have a larger effect on the deformation than in larger samples. Small pillars exhibit larger scatter in their stress strain data because the orientation of a single dislocation with respect to the loading direction determines the onset of plasticity. In larger pillars, there are always enough dislocations available, so that plasticity starts within a small intervall of stress. The increase in scatter with decreased sample dimensions is also observed in experiments [6, 9]. The power law exponent that was found in the study by Senger et al. was -0.57, which is in good accordance to what is observed in the experiments performed on fcc metals. Senger et al. observe three different mechanisms that lead to the mechanical size effect.

i) the initial distribution / statistics of the Frank-Read sources
ii) blocking of dislocation motion due to reactions and boundary constraints
iii) formation of new dislocation sources due to glissile junctions and cross slip

The latter mechanism leads to larger dislocation sources in larger pillars, which reduces the flow stress because larger sources need lower shear stresses in order to be activated.

Rao et al. simulated tetragonal and cuboid Ni samples with the DDD-technique [70]. The dimensions of their samples ranged from 1µm to 20µm where deformation in the smallest samples is controlled by single dislocations and deformation in the large samples is a matter of interacting dislocation networks. For the small samples the simulations show the effect of source truncation hardening and stress strain curves which show staircase-like hardening behaviour, as it is observed in the experiments. Furthermore strain hardening due to a lack of dislocations was observed, referred to as 'exhaustion hardening'. The simulation data of the smallest samples showed strong scatter which is attributed to the initial distribution of the Frank-Read sources. The stress strain curves get smoother and show less strain hardening as the dimensions of the simulation cell are increased. The initial dislocation source density, that is used for the simulation, affects the power law exponent of the mechanical size effect. The power law exponent varies from -0.43 ($\rho = 10^{13}\mathrm{m}^{-2}$) to -0.84 ($\rho = 7 \cdot 10^{11}\mathrm{m}^{-2}$).

A different theoretical approach is provided by Parthasarathy et al. [71] who created a mathematical model which describes the mechanical size effect of fcc metals in terms of the stochastics of dislocation source lengths in confined volumes. In samples that are small enough, which means that the sample dimension is on the order of the source length, the initially existing Frank-Read sources will transform into single arm sources. The model assumes a random distribution of pinning points, with each point being the starting point of a dislocation arm ending at the surface. The lower bound of the critical resolved shear stress (CRSS) is then given by the longest distance between a pinning point and the surface. The CRSS is determined according to

$$CRSS = \frac{\alpha\mu b}{\overline{\lambda}} + \tau_0 + 0.5\mu b\sqrt{\rho_{tot}} \tag{2.18}$$

with α being a geometrical constant, $\overline{\lambda}$ the mean value of source length, μ the shear modulus, b the Burgers vector, τ_0 the friction stress and ρ_{tot} the total dislocation density. $\overline{\lambda}$ depends on the sample size and therefore determines the size dependent behaviour. Equation (2.18) was used to simulate the mechanical size effect of Au and Ni. The results of these calculations are in good accordance with the experiments [6, 9, 49, 52]. The influence of single arm dislocations on the CRSS was also investigated by means of discrete dislocation simulations [60].

The simulations of fcc metals lead to a deeper understanding of the involved dislocation mechanisms. The simulations indicate that source truncation and dislocation starvation are important mechanisms that may lead to the size dependent behaviour in the experiments. A more detailed description of possible mechanisms that lead to size effects can be found in chapter (2.4.3). In general, the scatter in the data of small pillars can be explained by the stochastics of the initial dislocation distribution.

Possible Mechanisms Causing Mechanical Size Effects

The experiments and simulations of the fcc metals suggest different dislocation mechanisms that can lead to a size dependent mechanical behaviour. The following mechanisms are a description of the frequently used explanations for the mechanical size effect.

Dislocation Starvation

Strain hardening in metals occurs by the interaction of dislocations among each other or with other defects. Impeding the motion of dislocations results in hardening of the material. This is true when dimensions are considered that are much larger than the typical length scale of dislocation networks. In submicron pillars a different mechanism may cause hardening, an effect known as dislocation starvation [51]. In these pillars the mobile dislocations have a higher probability of annihilating at a nearby surface than of multiplying and being pinned by other dislocations. In this state, plasticity is determined by nucleation and motion of dislocations, dislocation interactions play a less important role. The dislocations leave the crystal faster than they can be multiplied, leading to a decreasing dislocation density. Dislocations are necessary for deformation, a lack of dislocations therefore leads to higher stresses. This mechanism will lead to a staircase-like stress strain response since activating or generating a dislocation source requires higher stresses than moving an existing dislocation.

Truncation of Frank-Read Sources

The stress to activate a dislocation source depends on the length of the source. In bulk samples, the critical resolved shear stress is largely determined by the stress required to initiate and maintain dislocation multiplication in the presence of a dislocation forest. The microplastic flow is known to initiate by multiplication of dislocations from the weakest source, typically a double-pinned Frank-Read source. The generated dislocations glide, interact with other dislocations and multiply by dislocation reactions. In samples of limited dimensions these multiplication processes are affected in the following manner.

First, upon operation the double-ended sources interact with free surfaces which results in truncated single arms of dislocations. The interaction of the single arm source with the surface is complex but it is a good approximation to assume that image stresses from the surface always tend to rotate the dislocation line until it is normal to the surface [72]. The stress required to activate these truncated sources depends on the shortest distance between the pinning point and the surface. This effect has been observed in large scale three-dimensional dislocation simulations [60]. This effect is controlling strength when the distance between internal obstacles (e.g. impurities, forest dislocations) is on the same order as the pillar dimensions when no obstacles are present in the pillars.

Second, the distance a dislocation can move is limited by the pillar size. Smaller pillars lead therefore to a reduced probability of cross-slip multiplication. These two mechanisms contribute to dislocation starvation, with a reduced number of mobile dislocations which can contribute to deformation.

If the pillar diameter is in the range of the mean source length, all Frank-Read sources will become single arm sources. The pinning points are randomly distributed in the sample and consequently there is a distribution of the source length of the single arm sources. The stress to initiate plastic strain is determined by the longest source which is the one which requires the lowest shear stress to be activated.

Dislocation Source-Limited Behaviour

According to the Hall-Petch relation (equation (2.9)), a large dislocation density (such as in a dislocation pile-up) leads to large internal stresses which can activate dislocation sources. Since the dislocation density is expected to be lower in small columns due to the loss of dislocations through the sample surface, it can be argued that higher stresses are required in small pillars to activate dislocation sources. Thus, large columns will have a high dislocation density and will readily reach the local stresses required to nucleate or activate sources. Discrete dislocation simulations show that this mechanism will lead to an inverse square root dependence of the yield stress on the pillar diameter [63].

Stochastical Size Effects

Mechanical size effects do not always have their origin in dislocation motion and interaction. In brittle materials, like ceramics, the fracture strength depends on the largest flaw in the considered volume. The probability of finding a flaw of critical size is larger

in large volumes than in small volumes. This means that on average small volumes show a higher fracture strength than large volumes. The failure probability of ceramics can be expressed in terms of extremal statistics e.g. using the Weibull distribution.
The weakest link approach can also be used to characterize microcompression experiments on metals. In a large pillar, there are many dislocation sources which are oriented statistically in all directions. If load is applied on a large pillar there will always be dislocation sources that can be easily activated. If the pillar diameter is decreased the number of dislocation sources decreases and the probability that at least one source can be easily activated decreases. In very small pillars there are only a few dislocation sources available, therefore the probability that a dislocation source can easily be activated is low. But if there is a source which is activated easily the yield strength of the pillar is low. This stochastic consideration of the mechanical size effect leads to the conclusion that yield strength and scatter of the data increases as the pillars become smaller. Discrete Dislocation Dynamics simulation confirm this behaviour [69].

Amorphization of the Surface

Pillar fabrication by Ga^+ ion sputtering will result in heavily damaged and Ga contaminated layers at the side walls of the columns. SRIM (Stopping Range of Ions in Matter) calculations show that the 30keV Ga^+ ions penetrate several tens of nanometers deep into the pillar material. In the case of Ta the mean penetration depth is approximately 10nm. The surface of the pillar is amorphized and acts as an obstacle for dislocation motion. The dislocations can pile up at the amorphized layer which results in strengthening of the sample. This mechanism corresponds to the one described in chapter (2.4.1) and therefore an inverse square root dependence of the yield stress on column diameter may be expected (cf. equation (2.9)).
If the amorphized layer was an obstacle for dislocation motion, irradiating with Ga^+ ions would lead to strengthening of the sample. Studies that compare irradiated to non-irradiated pillars suggest that Ga^+ contamination rather leads to softening than to strengthening since defects are introduced into the material [59, 73]. However these studies were performed on initially dislocation-free pillars. It is not clear how Ga^+ contamination affects the deformation of pillars having a significant initial dislocation density.
An inherent feature of focused ion beam machining is the contamination of the sample with Ga. Ga seems to affect the mechanical properties of the surface of the sample. Nanoindentation experiments performed on molybdenum-alloy single crystals show a hardening of the surface after the material was exposed to a Ga^+-beam [74]. Measure-

ments investigating the influence of Ga on Ta were carried out in this work (chapter (4.1)).

HCP Metals

Compared to the vast number of microcompression experiments made on fcc metals, size dependent mechanical behaviour of hexagonal close-packed (hcp) materials has hardly gained attention. Byer et al. performed microcompression on ⟨0001⟩-oriented magnesium single crystals [75], which exhibits a fundamentally different behaviour compared to fcc metals. In this study pillars with a diameter of 2.5μm and 10μm were tested and compared to tests on bulk samples obtained from [76] and [77]. It is noteworthy that in the investigated range no size effect was observed in magnesium. The 2.5μm pillars exhibited the same strength and hardening behaviour as the 10μm samples.
A study by Q. Yu et al. [78] on an hcp Ti-Al alloy exhibits a size dependent behaviour which is attributed to deformation twinning. In pillars with diameters smaller than 1μm deformation twinning is entirely replaced by ordinary dislocation plasticity and the mechanical size effect disappears.

BCC Metals

Not many microcompression studies of bcc metals have been carried out and therefore the knowledge about size-dependent behaviour of bcc metals is limited. The bcc studies that are available [10, 11, 79, 80] show a pronounced size effect in bcc pillars. The stress strain curves obtained from pillar compression do not show any qualitative differences between fcc and bcc pillars. Large pillars deform in a bulk-like manner whereas smaller samples show a stochastic deformation behaviour with pronounced strain bursts. As observed for fcc metals these strain bursts appear randomly with an arbitrary magnitude.
Bei et al. [59] do not see a size effect in their study on Mo-alloy micropillars. Their samples consisted of Mo-alloy fibres in a Ni matrix. The pillars were machined by etching the matrix, this means that FIB machining was not necessary and therefore the authors assume that the pillars have a low density of dislocations and other defects. It is argued that the non-existing size effect is rather an effect of the lack of defects than a property of the material. By prestraining the pillars, defects were introduced and consequently a size effect appeared.
If the size dependency of the bcc metals is characterized by a power law according to equation (2.16), it can be seen that bcc metals do not exhibit a uniform power law

exponent as it was observed for the fcc metals. The study of Nb of Kim et al. [11] reveals an exponent that is more than twice as large as the one found by Schneider et al. [80]. The power law exponent of the study by Kim et al. is based on only a few measurements, where data of compression experiments and tensile experiments were considered together. The microcompression experiments are subject to a certain error and bcc metals usually show a tension compression asymmetry, therefore the power law exponent of this study has to be considered with care.
The differences between the size dependent behaviour of fcc and bcc metals are discussed in more detail in chapter (6.3). Furthermore comparing different bcc metals may shed light on the underlying dislocation mechanisms that lead to the observed non-uniform mechanical size effect (cf. chapter (6.4)).

Simulations of Small Scale Plastic Deformation of BCC Metals

DD simulations of bcc metals are quite complex, due to the special structure of screw dislocations, the numerous glide systems and the non-Schmid behaviour of these materials. The particular mechanical properties of bcc metals are discussed in chapter (2.3).
In a study by Greer et al. [34] the evolution of a single dislocation in an fcc and bcc micropillar was simulated. In the fcc lattice, dislocations split into two partial dislocations. This dislocation configuration is not able to cross slip. On the contrary, the dislocation in the bcc lattice is not limited to move only on one glide plane. The DD-simulation also takes image stresses into account which affect the dislocation in the vicinity of the sample surface. In the fcc lattice under an applied shear stress the dislocation moves through the pillar without any interaction, leaving a dislocation starved sample behind. Applying a shear stress in the bcc pillar will turn the dislocation into a pure screw dislocation. The screw moves through the pillar and forms a cusp when it is close to the surface. The cusp formation can be attributed to image stresses (cf. chapters (6.5.2) and (6.6)) and leaves three dislocations behind, with on moving in the opposite direction of the other two dislocations. These two screw dislocations escape through the surface, whereas the cusp formation and self multiplication occurs again (figure (2.7)). These simulations suggest that dislocation starvation is unlikely to occur in bcc metals because of the self multiplication of the screw dislocations.

Microcompression of Cuboidally Shaped Tungsten Pillars

Most of the microcompression studies were performed on cylindrical pillars. This pillar shape was chosen for investigating size effects since it provides the simplest stress state. Cuboidal pillars show stress concentrations at the corners and therefore the stress state in these samples is more complex. For the determination of specific material properties the cuboidal shape can be of advantage. A study by N.M. Jennett et al. [81] deals with a phenomenon called 'The thinness effect'. Cuboidal pillars made of tungsten were tested. By varying the ratio of long axis to short axis it can be investigated which characteristic length determines the size dependent behaviour. From this investigation can be concluded that the flow stress of the pillar is determined by the shortest dimension of the pillar.

Another application for cuboidal pillars is the investigation of deformation behaviour with respect to the in-plane orientation of the metal. This technique can give new insights into dislocation processes and was used in this work. A detailed description of this technique can be found in chapter (3.5).

3

Experimental

3.1 Dual Beam SEM and FIB

The microcompression samples used in this work were exclusively prepared by Focused Ion Beam (FIB) machining. For milling and imaging a Dual Beam FIB (FEI Nova Nanolab 200, FEI Hillsboro, Oregon 97124, USA) was used. Besides an SEM column which is mounted vertically, it is also equipped with a Ga^+ column which is mounted at an angle of 52° with respect to the SEM column (figure (3.1)). The electron beam is generated by a field emission gun (FEG) and subsequently the electrons are accelerated by an acceleration voltage which is on the order of several kV. Usually, electron beam images were taken at 10kV. The maximum resolution of the SEM is approximately 1nm. The second column emits Ga^+ ions which are accelerated by a voltage of 30kV. The Ga^+ beam can be either used for imaging or milling. Its maximum resolution is approximately 7nm. Milling occurs by the impact of the Ga^+ ions on the sample surface. Consequently, the sputter rate of the ion beam depends on the acceleration voltage, the current and the sample material. The resolution of the ion beam decreases with increasing beam current. The beam current can be adjusted in a range from 10pA to 20nA. Exposing the sample to the Ga^+ beam is always associated with removal of the sample material.

3.1.1 EBSD

The Dual Beam FIB does not only provide the possibility of imaging and milling, it can also be used for characterizing the crystallographic orientation of the sample material. This characterization was done by using Electron Backscatter Diffraction (EBSD). The system used for this work was an EBSD-detector manufactured by HKL (now Oxford Instruments, Tubney Woods, Abingdon, Oxfordshire, OX13 5QX, UK).

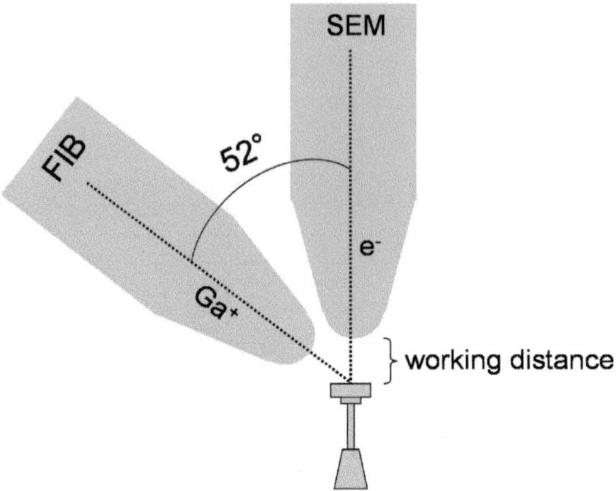

Figure 3.1: Schematic of sample and columns of the dual beam microscope

The mechanical properties of an individual crystallite are determined by its crystallographic orientation. EBSD is a useful tool to investigate textures, grain sizes and misorientations between individual crystallites. From an EBSD-measurement, the full 3D orientation information of a crystallite can be obtained in terms of three Euler angles.

The measurement procedure includes the following steps: The sample is tilted 70° to the horizontal and irradiated with a 20kV electron beam. The electrons interact with the atomic lattice planes of the crystalline structures and the backscattered electrons interfere with each other. If the Bragg condition is fulfilled, intensity in a backscatter diffraction pattern (Kikuchi pattern) can be observed. The Kikuchi patterns are characteristic for each crystal structure and orientation. The intersection of several Kikuchi bands represents a zone axis in the crystal lattice. This intersection has the same symmetry as the corresponding zone axis. For instance, a $\langle 111 \rangle$ direction in a cubic lattice shows three-fold symmetry and therefore the Kikuchi pattern shows a three-fold symmetry as well. The back-scattered electrons are detected by the phosphor screen of the EBSD detector and cause the screen to fluoresce. This signal is detected by a CCD camera and indexing the patterns is done with the help of commercial software (Channel 5) supplied by HKL.

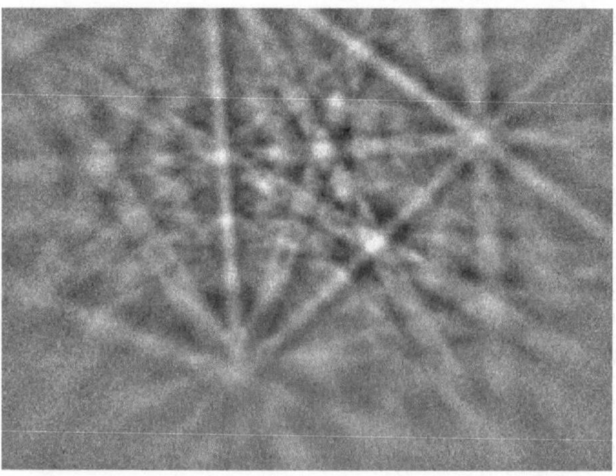

Figure 3.2: Kikuchi pattern from an Fe sample. The pattern is characteristic for the material and the crystal orientation. It shows several zone axes, for example a $\langle 100 \rangle$ zone axis with four-fold symmetry on the upper right.

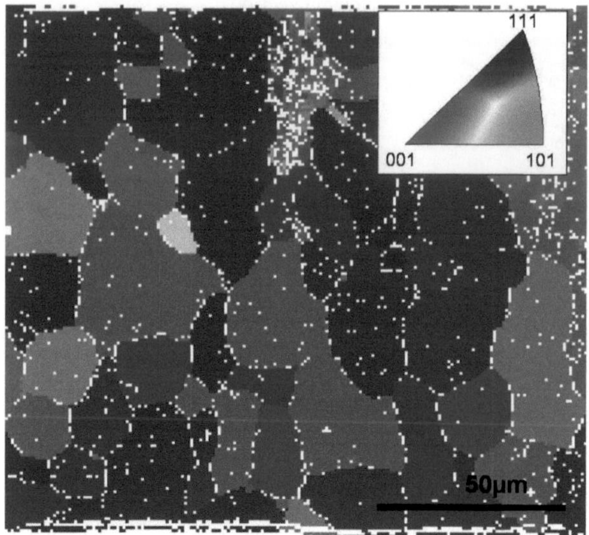

Figure 3.3: EBSD map for the out-of-plane orientation of a Ta sample

3.2 Sample Fabrication

Sample fabrication with the FIB can be done in two different ways, top-down milling or the more complex lathe method.

3.2.1 Top-Down Milling

Top-down milling is the most commonly used technique for pillar machining because it is comparatively easy and fast. The milling time depends on the material, the pillar size, the ion beam current and voltage; usual milling times for one pillar are in the range of 10-40 minutes. The top-down method consists of two steps: First, a coarse shape of the pillar is milled. Usually, for the coarse cut a current between 1nA and 20nA is used. The second step is the fine cut which gives the pillar its final shape. The fine cut is a single step milling process, which means that the trajectory of the ion beam has the shape of a spiral which passes through from the outside to the inside. The current of the fine cut is much lower than that of the coarse cut, common values are between 10pA and 0.5nA, depending on pillar size. The disadvantage of this method is, that the pillars do not have a perfect shape. The pillars show taper of up to 5° (figure (3.4)). A better geometry can be accomplished by the lathe-cut method, which is described in more detail in the following section.

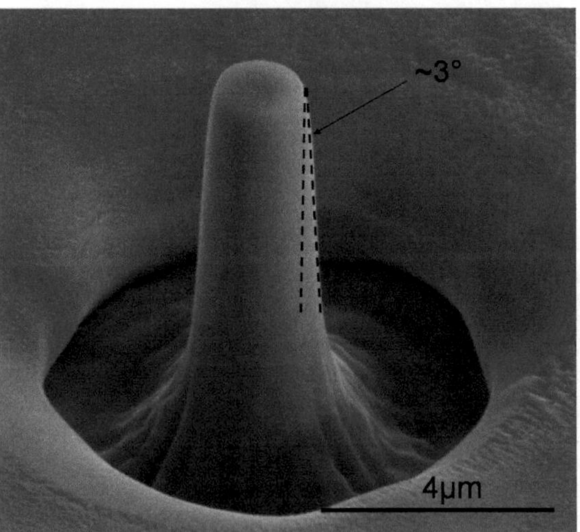

Figure 3.4: SEM micrograph of a Fe-pillar machined with the top-down technique. These pillars exhibit a taper angle of up to $\approx 5°$

3.2.2 The Lathe Technique

The sidewall taper of the pillars leads to a non-uniaxial stress state when the pillars are compressed. For a quasi-uniaxial stress state within the pillars it is necessary to optimize the shape of the pillars. The lathe method offers a technique that allows the fabrication of almost perfect compression samples with well defined dimensions. The lathe method was developed by Uchic et al. [6].

The lathe-cut itself is a multiple step process and can be automatised by using a scripting language (FEI Runscript) on the microscope. Before the lathe-cut method can be applied a coarse cut of the pillar has to be machined in the conventional way. On top of the pillar a shallow circle is milled which serves as a reference point for image recognition. The stage of the microscope is tilted to $-10°$, this means the incidence angle of the ion beam is $28°$ with respect to the sample surface. A pattern is milled at the edge of the pillar (the pattern has the shape of a flipped L) and subsequently the stage is rotated incrementally by a small angle. This process of milling and rotating continues until the sample has been rotated by at least $360°$. The shape of the pillars get better with small rotation steps, but this will increase the number of steps and therefore the milling time. Usually the angle increment is $6°$ and 66 milling steps are performed leading to milling times of $\approx 3-4$hours. The pillars cut by this technique do not show any sidewall taper (figure(3.5)), but the method is quite complex and time consuming and therefore was rarely used. A comparison between lathe and top-down pillars shows that there are no significant differences in their stress strain response for single crystal fcc samples.

For large pillars (diameter $\geq 10\mu$m) the coarse cut cannot be milled with the FIB within reasonable time. In this case electrical discharge machining (EDM) (cf. Appendix (A.1)) has been used to perform the coarse cut.

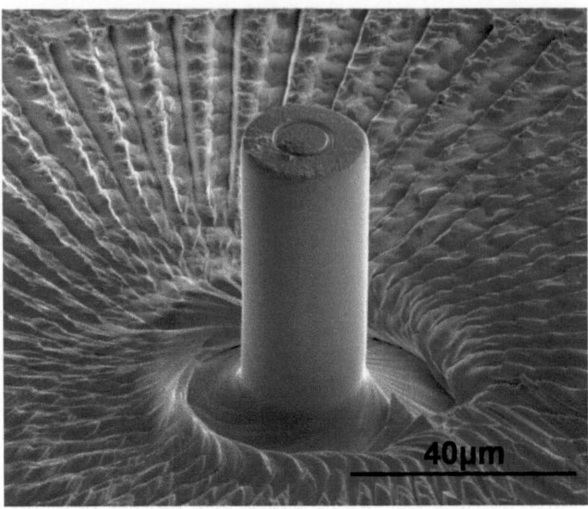

Figure 3.5: SEM micrograph of a Ta-pillar machined with the lathe-cut method. The coarse cut for this pillar was done by EDM.

3.3 Nanoindenter

The pillars machined by the FIB were compressed with a Nanoindenter XP (MTS Systems Corp., Eden Prairie, MN, USA). A schematic assembly of the nanoindenter is shown in figure (3.6). The indenter tip is mounted at the end of the indenter shaft. Depending on the experiment, different tip geometries are used, the most common are Berkovich indenters, spherical indenters, cube corners and flat punches. The flat punches have the shape of a truncated cone and are used for the microcompression experiments. Three different flat punches, having cross sectional diameters of 10µm, 20µm or 100µm, were used for the experiments of this work. For conventional indentation experiments a Berkovich tip was used which is a three sided pyramid with an opening angle of 142.3°. This tip has become the standard tip for modulus and hardness measurements [82] since four sided pyramids like the Vickers indenter cannot be easily machined at this size scale. Machining a sharp tip on a four sided indenter can result in a ridge.

However, a real Berkovich indenter is not ideally sharp which affects the force-displacement response of the material. For an ideal Berkovich indenter the projected area A with respect to indentation depth h can be calculated according to:

$$A = 24.5 \cdot h^2 \tag{3.1}$$

For a blunted tip the projected area A is fitted with a polynomial whose coefficients can be determined with the commercial software 'Analyst' (MTS Systems Corp., Eden Prairie, MN, USA).

Driving a current through the coil at the upper end of the indenter shaft results in a repulsive force between the permanent magnet and the coil. The load on the sample is controlled by the current which is applied. The spring support keeps the indenter shaft in position. The displacement of the indenter tip is measured by a three plate capacitor. Controlling the force by the applied current makes the nanoindenter an inherently load controlled device.

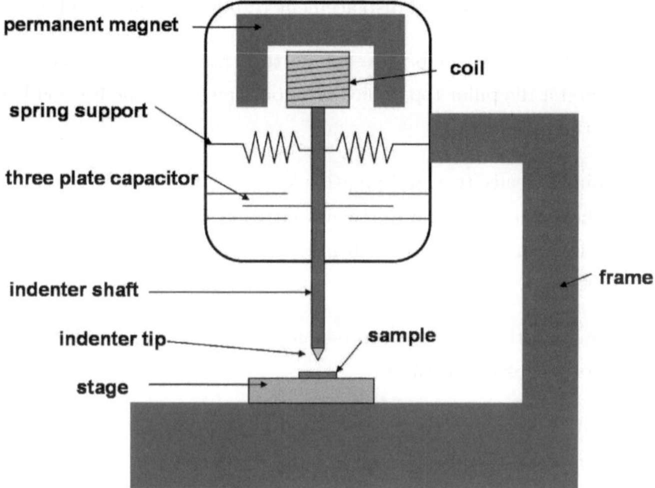

Figure 3.6: schematic assembly of the Nanoindenter XP (MTS)

3.4 Compression of Pillars

A flattened diamond tip, with a 10μm flat punch, was used for most of the experiments. For pillars larger than 4μm, tips with a punch diameter of 20μm or 100μm were used. All experiments were executed under load control which means that an increasing force is applied. Tests were stopped when the indentation depth reached 10% of the original pillar height. The pillars were compressed without any unloading steps in between. The loading rate was set in a way that a constant stress rate was effective. The stress rate was determined by using the loading rate and the initial cross section of the pillar.

Common values in the experiments of this work were between $5\frac{\text{MPa}}{\text{s}}$ and $50\frac{\text{MPa}}{\text{s}}$ aside from the strain rate sensitivity measurements, in which the load rate was varied over 3 orders of magnitude.
Although microcompression is a commonly used technique, it has some limitations that will be discussed in the following.

- The stress in the pillar is not perfectly homogeneous and uniaxial due to several reasons. In particular, the non-perfect geometry of the pillars leads to deviations from the uniaxial stress state.
- The pillars are connected to the substrate which means that the compliance of the substrate affects the load displacement response of the samples.
- Stress concentrations will be present at the transition from the pillar to the substrate and a the pillar top, which is a consequence of the friction between the pillar and the indenter tip.
- FIB machining leads to contamination of the sample surface with Ga^{+} ions. To date, the influence of the this contamination is not unambiguous, but there are indications that the sample surface hardens when it is exposed to an ion bombardment (cf. chapter (4.1)).

Besides these limitations, microcompression also offers also a number of advantages as compared to other testing methods:

- Microcompression allows the investigation of very small samples down to pillar diameters of about 150nm.
- Material properties can be obtained in a macroscopically almost non-destructive experiment.
- Sample preparation is quite simple and a large number of samples can be machined within reasonable time (cf. chapter (3.2.1)), leading to good statistics.
- The samples have a large surface to volume ratio and the investigated metals have a high heat conductivity, so that heating of the samples can be neglected.
- Furthermore microcompression experiments offer the opportunity to investigate the deformation morphology of the pillars, for instance active glide systems can be identified.
- The tests are site specific and individual grains can be investigated.

- It is possible to perform experiments on every material that can be machined by a focused ion beam.

3.5 Aligning Cuboidally Shaped Pillars with Respect to the Burgers Vector

Besides applying the microcompression technique to cylindrical pillars, microcompression was also applied to cuboidally shaped pillars. The orientation perpendicular to the pillar axis (referred to as in-plane orientation) was of great importance in these experiments. Therefore the alignment procedure is discussed in detail.
Cuboidally shaped pillars can be used in order to investigate the deformation behaviour with respect to the in-plane orientation of the sample. The orientation of individual crystallites can be expressed in terms of Euler angles. The Euler angles allow vectors to be transformed from the SEM-chamber system into the crystal coordinate system. Transferring coordinates from one coordinate system to another is done by rotating the initial coordinate system around its Z-axis using the Euler angle E1. Subsequently the resulting coordinate system is rotated around the X-axis using E2. A third rotation around the Z-axis with E3 completes the transformation procedure. The Euler angles allow a very accurate determination of the out-of-plane orientation (which is the loading direction) of the sample. The Schmid factors of all possible glide systems (48 for bcc metals and 12 for fcc) can be calculated according to equation (2.1).
The glide system with the highest Schmid factor is the one which is most likely activated when the pillar is deformed, since on this glide system the highest resolved shear stress is active. It has to be taken into account that Schmid's law just bases on geometrical considerations, other contributions like differences in the packing density of the glide planes are not considered. Glide in fcc metals obeys Schmid's law since all the $\{111\}$-planes are congeneric. The behaviour of bcc metals is more complex, since glide can occur on glide planes of different types. This leads to the often observed non-Schmid behaviour of bcc metals [24]. Nevertheless Schmid factors were used as first order approximation to try to align the pillars relative to the expected deformation direction.

The Burgers vector of the glide system with the highest Schmid factor is used to align the pillars. The long face of the cuboidal pillar was either parallel (long screw orientation, in the following referred to as LS configuration) or perpendicular (short screw orientation, in the following referred to as SS configuration) to the projection of the Burgers vector onto the sample surface. Figure (3.7) shows a sketch of the alignment of a cuboidal pillar in LS orientation. An SS oriented pillar is rotated by 90° around pillar axis. SEM-images of a LS and an SS oriented pillar are shown in figure (3.8).

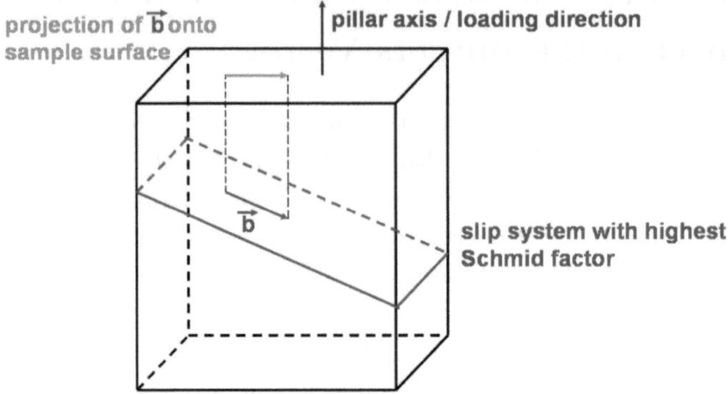

Figure 3.7: Alignment procedure of a cuboidal pillar in LS orientation.

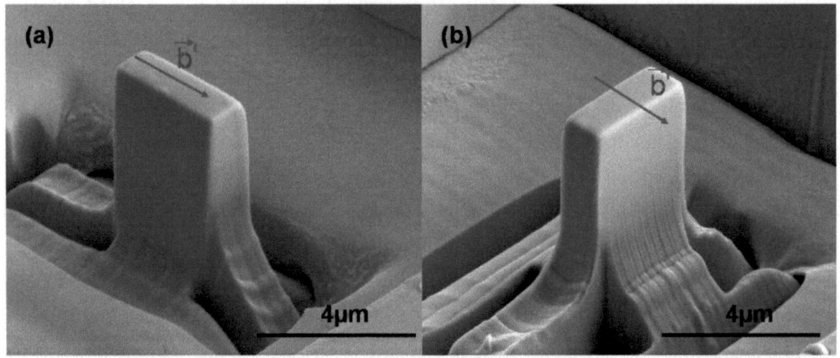

Figure 3.8: SEM micrographs: Cuboidal pillars aligned with respect to projection of the Burgers vector $\vec{b'}$ onto the sample surface. (a) LS configuration, (b) SS configuration

It is necessary to know which glide system will be activated during compression, that means that the out-of-plane orientation of the sample cannot be chosen arbitrarily. Single slip orientations work well since the difference between the highest two Schmid factors is maximal. In bcc metals another possibility is to use crystals with a $\langle 110 \rangle$ out-of-plane orientation [83]. In these samples only glide systems with a $[111]$ and $[11\bar{1}]$ Burgers vector have a Schmid factor different from zero. These two vectors span

a plane which is used for the alignment of the pillars. In this case LS configuration means that the long face of the cuboid is parallel to this plane, SS configuration denotes the perpendicular configuration. Orientations with multiple slip work as well, as long as the activated glide systems have the same Burgers vector.
Cuboidal pillar studies were performed on the bcc metals Ta and α-Fe and for comparison also on Cu, which has an fcc crystal structure.

For Cu and Ta, two different geometries (aspect ratios) were tested, the geometries are shown in figure (3.9). An explanation why it was necessary to perform the tests on two different geometries can be found in chapter (5.1.2). In the following the long geometry will be denoted as geometry 1, the short geometry as geometry 2. The samples machined in Fe were only tested in geometry 1. The pillars were compressed to a indentation depth of approximately 10% of their height. The results of the microcompression experiments on the cuboidally shaped pillars are given in the chapters (5.1.2), (5.2.3) and (5.3.3).

Figure 3.9: SEM micrographs: Geometry 1 (a) and geometry 2 (b) of the rectangular pillars.

3.6 Identification Procedure of Glide Planes

Microcompression experiments in combination with EBSD allow the identification of activated glide planes. The normal vector of the activated glide plane can be determined by using the crystallographic orientation of the sample and simple geometric considerations. This technique works for cylindrical pillars and for pillars which have a square cross section.
Glide plane identification for cylindrical pillars is slightly more complex than for pillars with a square cross section (cf. figure (3.10)). The glide plane is fitted with an ellipse

(black dotted). A rectangle (red) is drawn from the intersection of the left edge of the pillar with the glide plane to the intersection of the right edge with the glide plane. The height of the red rectangle is a measure for the inclination angle of the glide plane in the image plane. The inclination angle of the glide plane perpendicular to the image plane can be determined as follows. The cross section of the pillar (blue ellipse) and the black dotted ellipse are tilted against each other in the out-of-image-plane direction. The green lines are a measure for this tilt angle, whereas the purple line represents the pillar diameter. The lengths in z-directions have to be corrected for tilt, since the image was taken at a tilt angle of 40°. The two inclination angles of the normal vector of the glide plane and the cylinder axis z of the pillar can be determined by the following relations,

$$\tan\alpha = \frac{h}{\sin 40° x} \tag{3.2}$$

$$\tan\beta = \frac{y_1 + y_2}{\sin 40°} \cdot \frac{\sin 40°}{d} = \frac{y_1 + y_2}{d} \tag{3.3}$$

with α inclination in image plane and β inclination perpendicular to image plane. y_1 and y_2 should have the same length, if the glide plane is perfectly fitted. The cylinder axis of the pillar is obtained from an EBSD measurement and represents the out-of-plane orientation of the sample. The normal vector of the plane has now to be transferred into the crystal system which is done by using a rotary matrix which contains the Euler angles. This procedure is quite complex and inaccurate, but it allows to extract the 3-dimensional information of the glide plane from only one image.

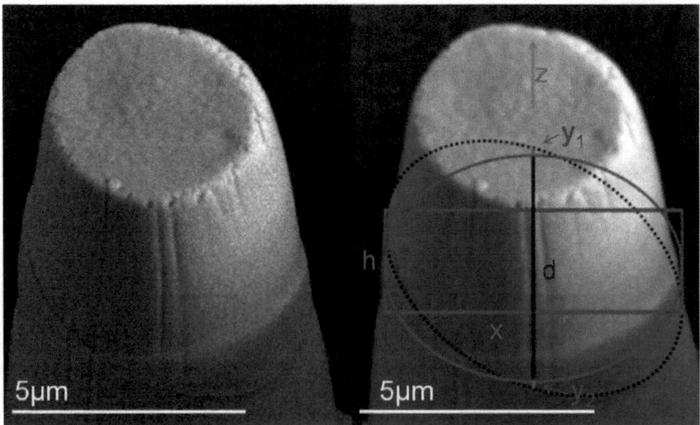

Figure 3.10: SEM micrographs: Glide plane identification of a cylindrical pillar. The images were taken at 40° tilt.

A more convenient way of identifying glide planes is using pillars with a square cross section (cf. figure (3.11)). This method requires two images of the deformed pillar in order to get the normal vector of the glide plane. The images are rotated by 90° against each other. Prior to compression the crystal orientation was measured with EBSD. The slip traces in the two images are fitted with the vectors $\vec{V1}$ and $\vec{V2}$. $\vec{V1}$ consists of the two components Δy and Δz, $\vec{V2}$ consists of Δx and Δz. The components Δz have to be corrected since the images are taken under 52° tilt, that means that $\Delta z_{true} = \frac{\Delta z}{\sin 52°}$. The normal vector $\vec{n}$ is obtained by

$$\vec{n} = \vec{V1} \times \vec{V2} \tag{3.4}$$

When the normal vector $\vec{n}$ is determined, it has to be transferred into the crystal system of the sample. Like mentioned above, this transfer procedure is done by applying the Euler rotary matrix on the vector $\vec{n}$.

Figure 3.11: SEM micrographs: Glide plane identification of a pillar with square cross section. The images were taken at 52° tilt.

3.7 Sample Materials and Specimens

Microcompression experiments were performed on Cu, Ta and α-Fe. The specifications of the materials like purity and crystallographic orientation are described in the following.

3.7.1 Copper

Cylindrical Pillars

A well annealed electropolished polycrystal (Alfa Aesar) with a purity of 99.9999% (metals basis) was used. The pillars were machined into a single grain of a Cu polycrystal with a diameter of approximately 700µm. All pillars were located in that specific grain having an out-of-plane orientation of ⟨235⟩, which is a single slip orientation. The highest Schmid factor exceeds the second highest by 24%. The pillar diameters were in a range between 225nm and 5µm, with the diameter measured at midheight of the pillar. The compression of the pillars was performed with a constant stress rate of $\approx 20\frac{\text{MPa}}{\text{s}}$ to a depth to approximately 10% of the pillar height.

Cuboidally Shaped Pillars

The cuboidally shaped pillars were milled into the same grain as the cylindrical pillars. The compression of the cuboidal pillars was performed in a similar way as compression of the cylindrical pillars, with a depth of approximately 10% of the pillar height and a constant stress rate of $\approx 20\frac{\text{MPa}}{\text{s}}$. The geometries that were tested had the following specifications (cf. figure (3.9)): $4.5 \cdot 1 \cdot 3\mu m^3$(height · width · length) and $4.5 \cdot 1 \cdot 5\mu m^3$(height · width · length)

3.7.2 Tantalum

Cylindrical Pillars

For the measurements of ⟨111⟩ and ⟨100⟩ oriented Ta (cf. chapter (5.2.1)) a Ta-sheet manufactured by Alfa Aesar was used. The sheet had a purity of 99.95% (metals basis) and was annealed in vacuum at 1000C° for one hour. The average grain size determined by EBSD was approximately 20µm. The sample was bonded to a stub by using conductive silver paint.
Before machining the pillars, a $150 \cdot 150\mu m^2$ area was irradiated with the 30kV Ga^+-ion beam for 12 minutes. The current (flux) of the beam was 20nA. This procedure removes the oxide layer from the sample which allows an EBSD measurement. From the EBSD map the grain orientations were obtained and grains with the desired orientation could be selected. The Ta-sheet was cold rolled so that it showed a texture. The crystallites showed predominantly a ⟨111⟩ out-of-plane orientation. For the preparation of the pillars ⟨111⟩ and ⟨100⟩ orientations were used. A maximum misorientation of 10° was tolerated. The pillars were machined into these grains with the top down method. To avoid stress concentrations in the pillars, the pillars were placed distant from grain boundaries.
Cylindrical pillars were also tested on a single crystal oriented for single slip. The single crystal with a purty of 99.99% was manufactured by MaTeck (52428 Juelich, Germany). The out-of-plane orientation of the crystal was close to [20 7 45], whereas areas with an orientation close to [3 2 5] were also found.

Cuboidally Shaped Pillars

The experiments with the cuboidally shaped pillars were also performed on the crystal oriented for single slip (out-of-plane orientation [20 7 45]). The glide system with the highest Schmid factor is $[1\bar{1}1]$ (011) and its difference to the second highest is 11.3% (for these considerations the {321} systems have been neglected since they are rarely observed). Tests were performed on two different geometries (cf. figure (3.9)), $4.5 \cdot 1 \cdot 3\mu m^3$(height · width · length) and $4.5 \cdot 1 \cdot 5\mu m^3$(height · width · length)

3.7.3 alpha-Iron

Cylindrical Pillars

For the experiments on Fe, a rod manufactured by Alfa Aesar was used. The metallic purity of the sample was 99.95 + %. The content of non-metallic impurities of the sample was analyzed by the Chemical Analytics Department of the Institute for Materials Research I, KIT. The results of this analysis are shown in table (3.1). The rod had a diameter of ≈ 1cm and a height of ≈ 6cm. The rod was cut into little discs with a height of several millimeters using a diamond wire saw. The discs were annealed for two hours at 850C°. After annealing the samples were grinded in several steps down to a surface roughness of ≈ 10μm. The grinding steps were followed by several polishing steps with diamond suspension down to a surface roughness of ≈ 1μm. For a last mechanical polishing step an oxide polishing suspension (OPS) was used. This abrasive contains SiO_2 particles which are only a fraction of a micrometer in size. All the grinding and polishing steps up to this point remove material mechanically and introduce therefore dislocations into the sample. The damaged surface layer was removed by electropolishing. In this polishing procedure the sample serves as an anode in an electrochemical cell. Here material is removed electrochemically and no mechanical defects are introduced into the sample. Furthermore the surface roughness is decreased since material removal is stronger at the elevated parts of the surface than at the lower parts. After these polishing steps the sample was annealed again for two hours at 850C° in order to remove damage and to allow for grain growth.

EBSD measurement and the FIB-preparation of the pillars is done analogically to the Ta sample. The EBSD-measurement revealed that the grain size was on the order of several hundred micrometers.

Element	Amount (mass-%)
C	<0.002
S	0.0017
O	0.0057 ± 0.0004
N	0.00021 ± 0.00001

Table 3.1: Chemical analysis of the used Fe sample

Cuboidally Shaped Pillars

The cuboidally shaped pillars were milled into the same polycrystal as the cylindrical pillars. A grain with an [122] out-of-plane orientation was chosen for the experiments on the cuboidal pillars. This is not a single slip orientation but the glide system having the five highest Schmidfactors all exhibited the same Burgers vector, namely $\left[\bar{1}11\right]$. The pillars were machined with the following dimensions $5 \cdot 1 \cdot 5 \mu m^3$(height $\cdot$ width $\cdot$ length).

4

Pre-studies

As has been mentioned in chapter (3.4), microcompression testing of FIB-machined pillars has some limitations. Therefore, the damage of the sample induced by Ga-irradiation and the effect of the substrate compliance have been investigated. Furthermore, it will be shown that microcompression experiments can be used in order to obtain the Young's modulus of a material, when the substrate compliance is taken into account properly.

4.1 Gallium Induced Damage

Irradiating a material with a 30keV Ga^+-ion beam will have an influence on the structure of the machined surface. Experiments show that irradiation may also change the mechanical behaviour of a material in terms of hardness and dislocation density. Compression experiments performed on non-irradiated Molybdenum-fibres do not show a size effect and show stresses in the range of their theoretical shear strength. This situation changes when these fibres are exposed to a Ga^+-ion beam. The compression experiments then show a size effect. Prestraining the fibres and compressing them subsequently leads to a size effect as well, so that it seems that the pillar needs some internal defects in order to show size dependent behaviour [59, 73, 74]. Transmission electron microscopy investigations confirm the damaging influence of FIB machining on the pillar.

Monte Carlo calculations using the software SRIM (stopping and range of ions in matter, www.SRIM.org) show that the mean penetration depth of 30keV Ga^+ ions into Ta is approximately 10nm when the incidence angle (angle between ion beam and sample surface) is 90° and the fraction of backscattered ions of the total number of ions was found to be 11.48% (figure (4.1)). The penetration depth of the ions follows a Maxwell-Boltzmann distribution, so that individual ions can penetrate much deeper than 10nm into the material. For small pillars with diameters of only several hundred nanometers the Ga contamination may change the mechanical behaviour. Top-down machining of the samples (cf. chapter (3.2.1)) may lead to less Ga contamination compared to the lathe-cut technique (cf. chapter (3.2.2)), since the lathe-cut technique requires more irradiation time.

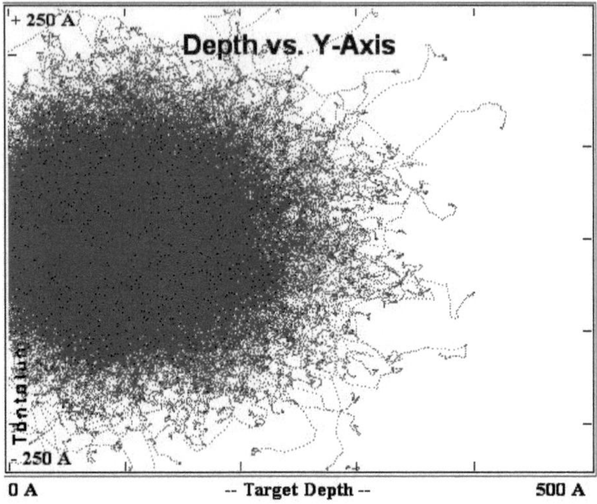

Figure 4.1: SRIM simulation of 30keV-Ga ions penetrating in Ta. The Ga^+ penetrate perpendicular into the Tantalum sample.

Conventional indentation experiments according to the Oliver and Pharr method [82] on Tantalum were performed in order to investigate the effect of Ga^+ irradiation. Four areas on the Tantalum sample were exposed to the same amount of irradiation but with different acceleration voltages. For comparison a non-irradiated area was tested as well. The results can be seen in figure (4.2) and confirm surface hardening of irradiated samples. The indentation depth was 500nm whereas the penetration depth of the ions is approximately 10nm as can be seen from figure (4.1), therefore the hardening effect in this measurement is underestimated.

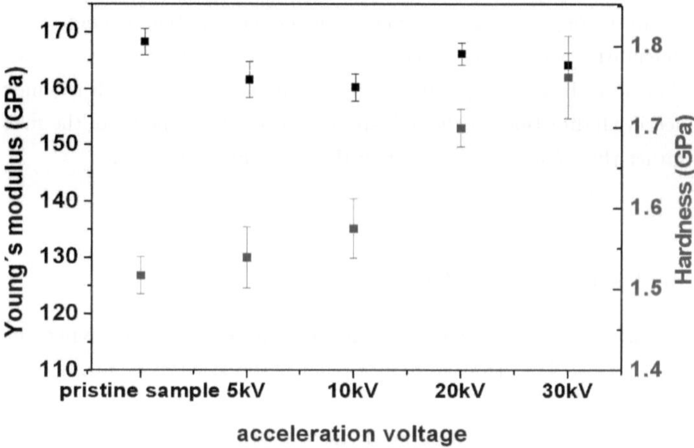

Figure 4.2: Indentation of an irradiated Ta sample. Young's modulus E and hardness H are plotted with respect to the acceleration voltage of the Ga ions. The error bars represent the standard deviation of the measurements.

4.2 Determination of Young's Modulus Using Microcompression Testing

4.2.1 Motivation

For a large number of applications it is neccessary to know the elastic properties of a material. In small scaled samples, Young's moduli can be measured by nanoindentation with a Berkovich tip. This procedure works quite well when the Young's modulus of the bulk material shall be investigated. Small samples as they are used in microcompression experiments usually consist only of a few crystallites or are single crystalline. In this case the Young's modulus depends on the orientation of the crystallites and measuring the elastic properties by nanoindentation will not work anymore. Nanoindentation with a Berkovich indenter creates a very complex stress state beneath the indenter. If a single grain is indented the material beneath the indenter will be strained in all directions of the half-space. The elastic response of the material is therefore rather that of a poly-crystal than that of a single grain [84].

4.2.2 The Experimental Concept

Microcompression offers a way to determine the orientation dependent Young's modulus of individual grains. The stress state in a pillar is supposed to be uniaxial, consequently it is simpler than the complex stress state beneath a Berkovich indenter in a conventional indentation experiment. Considering the stress state, a tensile experiment should be preferred, but performing these experiments at the micrometer size scale is experimentally challenging. As mentioned in chapter (3.4) the microcompression technique is subject to several limitations. The geometry of the pillars is not ideal in terms of taper, furthermore the pillars are attached to the underlying material which has a certain compliance. These limitations will lead to an apparent reduced modulus in the experiment. The reduction in modulus depends on the height to diameter ratio of the pillar. A study consisting of Finite-Element Simulations and microcompression experiments was carried out in order to determine the orientation dependent modulus of a single crystalline material. Silicon (Si) is an appropriate material for a first proof of concept, since it shows an anisotropy of its elastic constants and behaves elastically until failure. This allows for a very precise determination of the elastic behaviour. Pillars were milled into a Si⟨111⟩- and a Si⟨100⟩-wafer having different aspect ratios between one and three. One pillar of each aspect ratio and orientation was compressed until failure. From these experiments the failure load respectively the failure stress was obtained. A representative pillar and the corresponding stress strain curve are shown in figure (4.3) and (4.4).

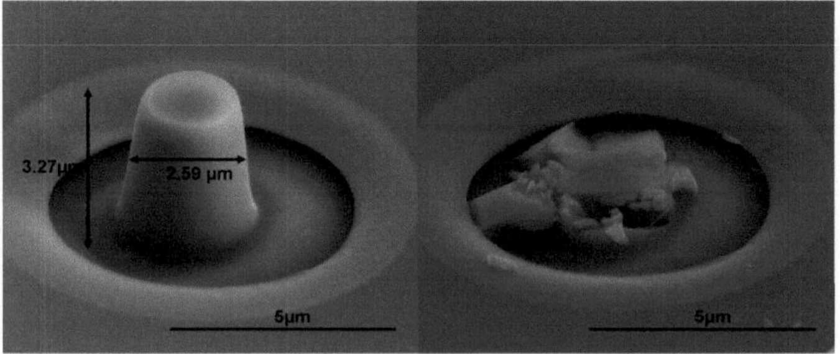

Figure 4.3: SEM micrographs: Si⟨100⟩ pillar compressed until failure.

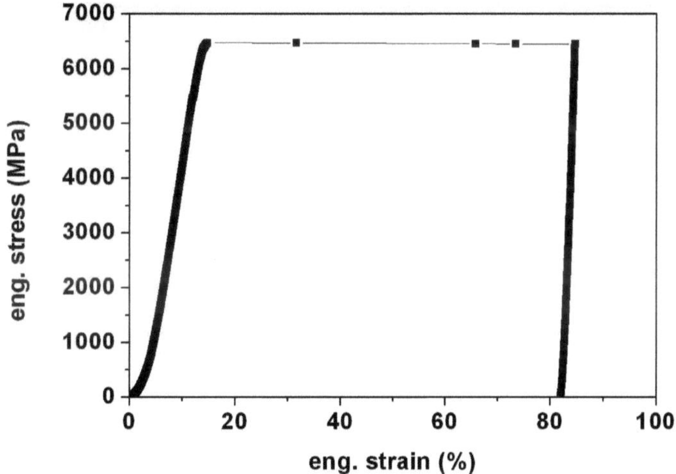

Figure 4.4: Stress strain curve of the pillar from figure (4.3)

The following experiments were conducted in the elastic regime. The pillars were loaded to a maximum load of $\approx$ 80% of the failure load with six unloading steps in between. A diagram of the loading-unloading cycles is shown in figure (4.5). The modulus was determined from the unloading segments. Since the unloading segments become non-linear when the load is close to zero, the modulus was calculated from the first 20% of the unloading segment. Every pillar was compressed with a loading rate of $250\frac{\mu N}{s}$ ($\approx 50\frac{MPa}{s}$). A representative stress-strain plot is shown in figure (4.6). It can be seen that the unloading curve does not exactly follow the loading curve, this means that a small amount of plasticity may be involved in the compression.

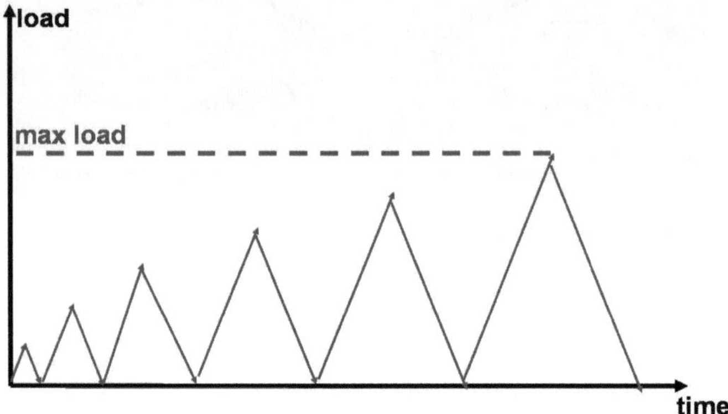

Figure 4.5: Diagram of the loading-unloading cycles of the Si pillars.

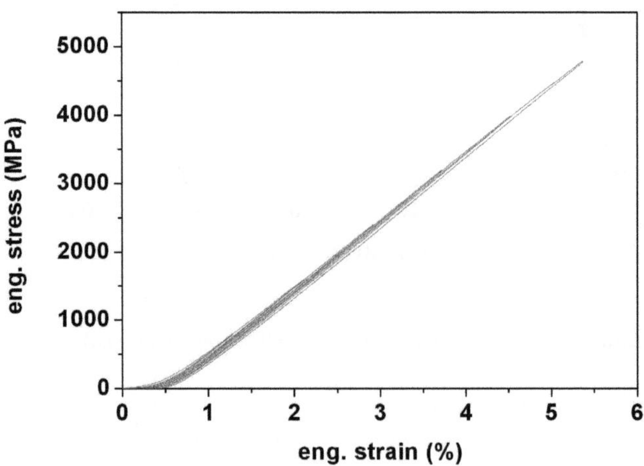

Figure 4.6: Compression of Si⟨100⟩ pillar with 6 unloading steps for modulus determination

For each sample, six values of the Young's modulus are obtained. The first and second value are usually smaller than the subsequent values where a saturation is reached. This might be due to early plasticity (small particles on the pillar or indenter) and misalignment of the pillar and the indenter. The modulus from each experiment is calculated by taking the arithmetic mean value of the last four modulus values. In figure (4.7) the modulus values of four Si⟨100⟩ pillars with an aspect ratio of ≈ 2.7 are plotted. The experiments for the Si⟨111⟩ were conducted in the same way.

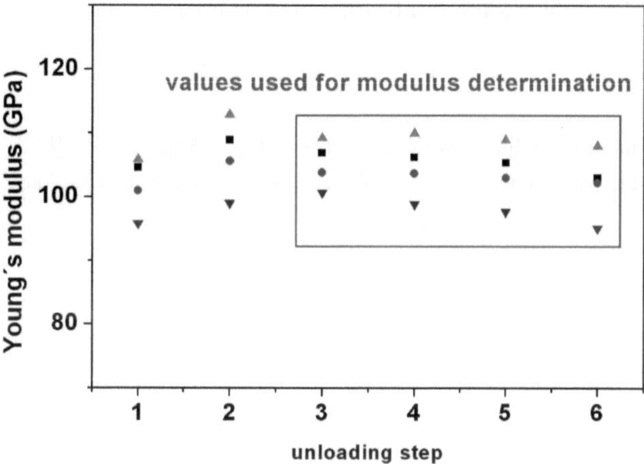

Figure 4.7: Moduli from six unloading steps of four representative Si⟨100⟩ pillars. The mean value of the last four values was used for comparison with simulation (values in the red box).

4.2.3 Results and Discussion

The values for the Young's modulus obtained from the experiments are lower than the theoretical value $E_{theo} = 130\text{GPa}$. This can be attributed to the substrate compliance. Finite element simulations were carried out in order to determine the influence of the substrate on the elastic behaviour of the pillar. Young's moduli for aspect ratios from one to nine and for $\langle 111 \rangle$ and $\langle 100 \rangle$ orientation were simulated. The taper of the pillars was also taken into account by the simulation. The Young's modulus increases as the aspect ratio of the pillars increases. This behaviour can be expected, since substrate compliance plays a less important role for a high aspect ratio than for a small one. As can be seen from figure (4.8) the limit for very high aspect ratios is the theoretical value of the Young's modulus which can be calculated by using the elastic constants.

$$\frac{1}{E_{hkl}} = S_{11} - [2(S_{11} - S_{12}) - S_{44}] \cdot \left(\alpha^2\beta^2 + \alpha^2\gamma^2 + \beta^2\gamma^2\right) \tag{4.1}$$

with E_{hkl} Young's modulus in hkl-direction, $\alpha = cos\left([hkl], [100]\right)$, $\beta = cos\left([hkl], [010]\right)$,$\gamma = cos\left([hkl], [001]\right)$ and compliances S_{ij}. For Si the following values can be found from literature [85].

$C_{ij} : 10^9\text{Pa}$ $S_{ij} : 10^{-12}\text{Pa}$	C_{11}	C_{12}	C_{44}	S_{11}	S_{12}	S_{44}
Si	165.7	63.9	79.6	7.68	-2.14	12.6

Table 4.1: elastic constants of Si

With the values from table (4.1) the Young's moduli for the $\langle 111 \rangle$-, and $\langle 100 \rangle$-orientation can be determined $\Rightarrow E_{111} = 189\text{GPa}$ and $E_{100} = 130\text{GPa}$. A graph of equation (4.1) is shown in figure (4.9) with the distance of the data point to the origin representing the Young's modulus of the particular direction.

The simulations result in two curves for each orientation. The first one simulates the modulus as it is obtained from the experiment (E111Ex respectively E100Ex). In the experiment displacement is measured at the pillar top containing the deformation of the pillar and the substrate compliance. The second curve in the simulation takes the substrate compliance into account by subtracting the substrate compliance ΔL_S from the total displacement ΔL_{Ex}.

$$E_{Ex} = \frac{FL_0}{A\Delta L_{Ex}} \tag{4.2}$$

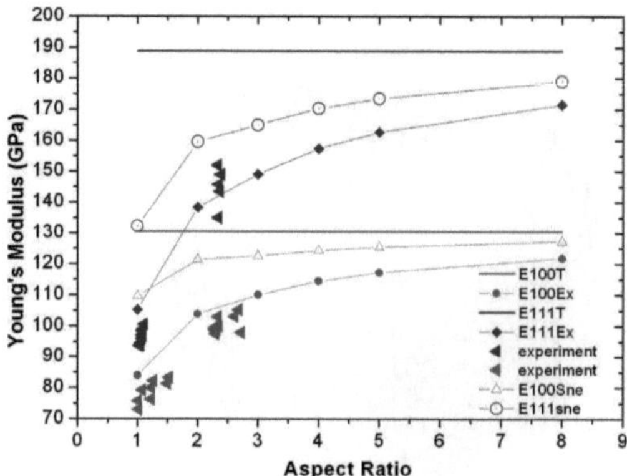

Figure 4.8: Simulation and results from experiment of the orientation dependent Young's modulus of Si. The labels have the following meanings: **E100T** ⇒ theoretical value of the Young's modulus in ⟨100⟩ direction, **E100Ex** ⇒ simulation of the Young's modulus in ⟨100⟩ direction without taking substrate compliance into account, **E100Sne** ⇒ simulation with Sneddon's correction for substrate compliance, **experiment** ⇒ data obtained from compression experiments. Plots for ⟨111⟩ direction are labelled in an analogous way.

$$E_{Sne} = \frac{F L_0}{A\left(\Delta L_{Ex} - \Delta L_S\right)} \tag{4.3}$$

$$\Delta L_S = \frac{1-\nu^2}{E}\frac{F}{2a} \tag{4.4}$$

F is the load, L_0 the initial length of the pillar and A the cross section of the pillar. The results of the experiments and the simulations are shown in figure (4.8) and it turns out that the experiments and simulations are in reasonable accordance with each other.

Microcompression can be used for the determination of the Young's modulus of particular crystallographic directions. The experimental data suggests that compression of pillars with a high aspect ratio will lead to a Young's modulus which is close to the theoretical value. The simulations show that the relation between Young's modulus E and aspect ratio ar can be determined according to

$$E(ar) = E_{theo} \cdot \left(1 - e^{-\frac{ar^b}{c}}\right) \tag{4.5}$$

with E_{theo} the theoretical value of the Young's modulus and b and c fit parameters. Simulations and experiments are in good accordance with each other, so that it is

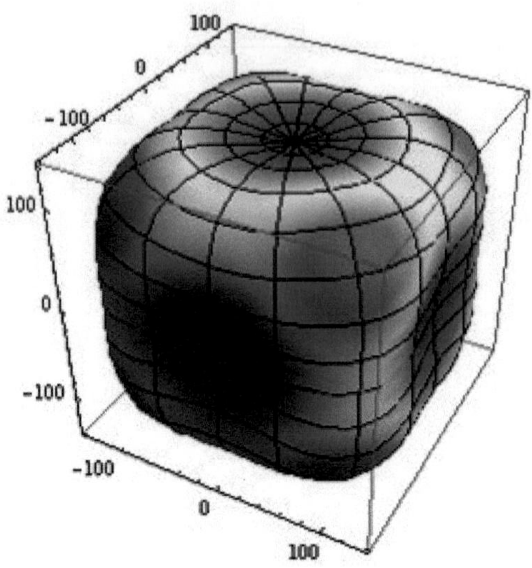

Figure 4.9: Orientation dependent Young's modulus of Si. The distance from the surface to the origin represents the modulus of the particular direction. The highest values are reached for the $\langle 111 \rangle$-directions (red areas) whereas the $\langle 100 \rangle$-directions exhibit the smallest values (blue areas).

possible to obtain the theoretical value of the Young's modulus from experiments in combination with equation (4.5).

Figure (4.10) shows the simulated and the measured data of Si$\langle 100 \rangle$. The simulated data points are fitted according to equation (4.5). It can be seen that this relationship represents the dependency of the Young's modulus with respect to the aspect ratio very well. Figure (4.11) shows that the fit procedure also works well for Si$\langle 111 \rangle$.

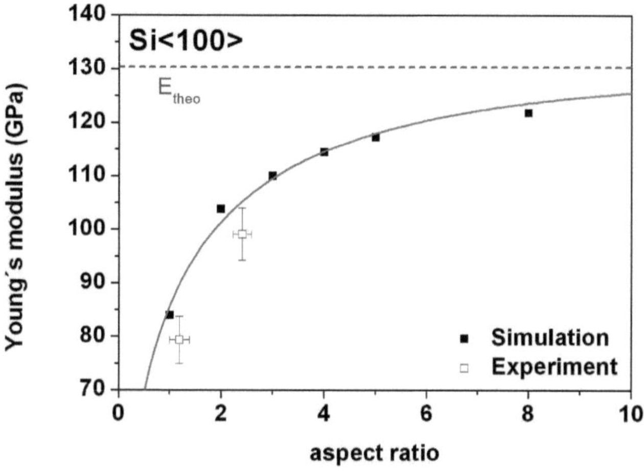

Figure 4.10: Simulated and experimental data of Si $\langle 100 \rangle$. A fit of the simulated data points according to equation (4.5) (green line) is added in the plot.

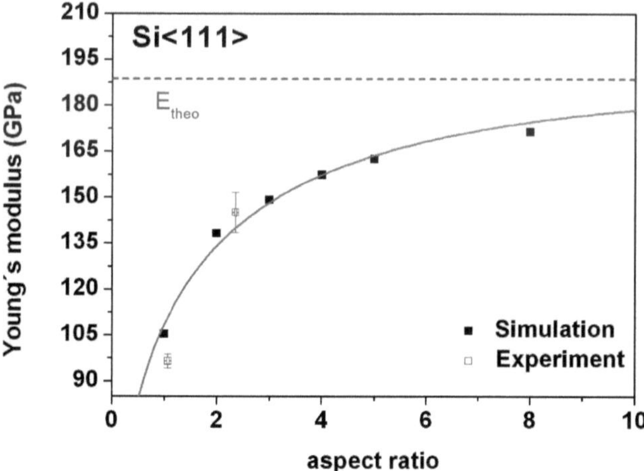

Figure 4.11: Simulated and experimental data of Si $\langle 111 \rangle$. A fit of the simulated data points according to equation (4.5) (green line) is added in the plot.

5

Results

5.1 Copper

5.1.1 Size Dependent Deformation

Microcompression studies of fcc metals reveal that mechanical size effects in these materials exhibit a uniform scaling behaviour. Microcompression experiments on Cu were performed in order to obtain reference data for a typcial fcc metal.

Cu shows a pronounced size effect where the larger samples deform in a bulk-like manner and the small pillars deform by stochastic events, as can be seen in figure (5.1). The investigation of the compressed pillars reveals that all pillars deformed by activating the same glide plane, namely $(1\bar{1}1)$(cf. figure (5.2)). A detailed description of the glide plane identification can be found in chapter (3.6). The flow stresses of the compression tests were measured at 5% plastic strain and plotted with respect to pillar diameter (figure (5.3)). The size effect is expected to follow a power law (cf. equation (2.16)) and therefore the data are plotted onto a double logarithmic scale. The data points follow a trend line whose slope corresponds to the power law exponent β. The power law exponent found in these experiments is -0.55 and agrees well to the value found by Volkert (β=-0.59) [8].

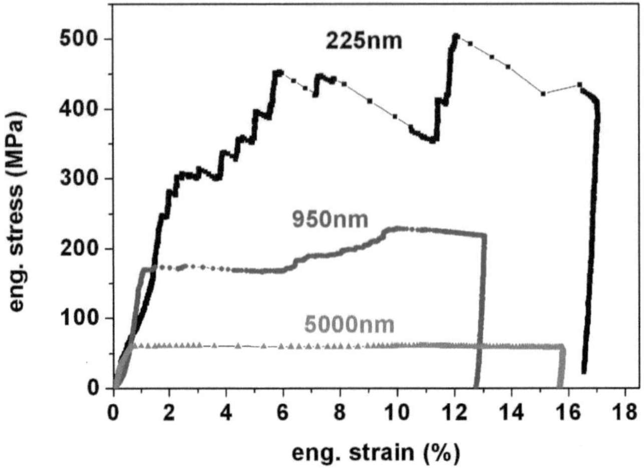

Figure 5.1: Representative stress strain curves from Cu microcompression. It can be clearly seen that there is a transition from bulk-like deformation behaviour to deformation behaviour exhibiting strain bursts (numbers give the diameter of the pillars at midheight).

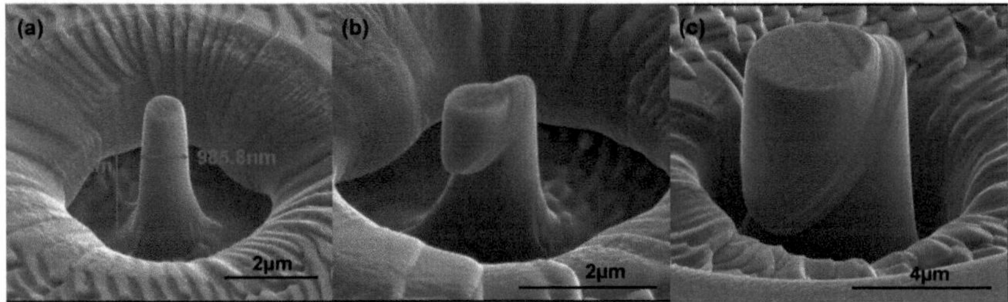

Figure 5.2: SEM micrographs: (a) Determining the dimensions of the pillars (b) compressed 1µm pillar (c) compressed 5µm pillar, (b) and (c) show glide on the same glide system.

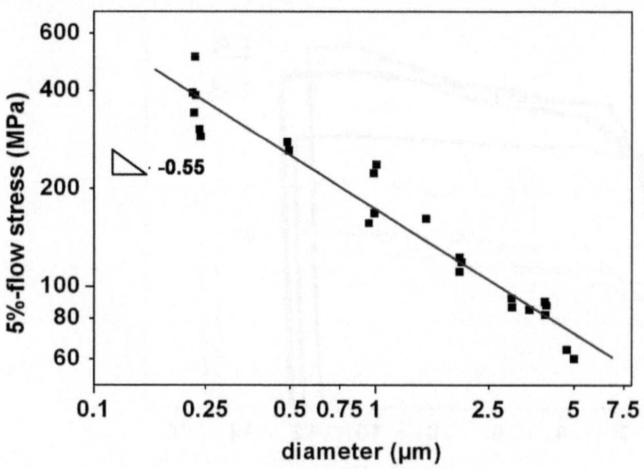

Figure 5.3: Flow stresses of cylindrical Cu pillars at 5% plastic strain plotted with respect to pillar diameter

5.1.2 Deformation of Cuboidal Cu Pillars

Geometry 1

The stress-strain curves of the Cu samples with geometry 1 (5:5:1 μm^3, H:L:W) show a clear orientation dependent behaviour, which cannot be expected in terms of screw dislocation mobility (figure (5.4)). An investigation of the compressed pillars revealed that higher stresses of the LS configuration originate from the substrate. The slip lines of the LS configuration run into the substrate whereas the slip lines in the SS configuration step out at the sample surface (figure (5.5)). To avoid this substrate effect the pillars were also tested with a different geometry (cf. chapter (3.5)).

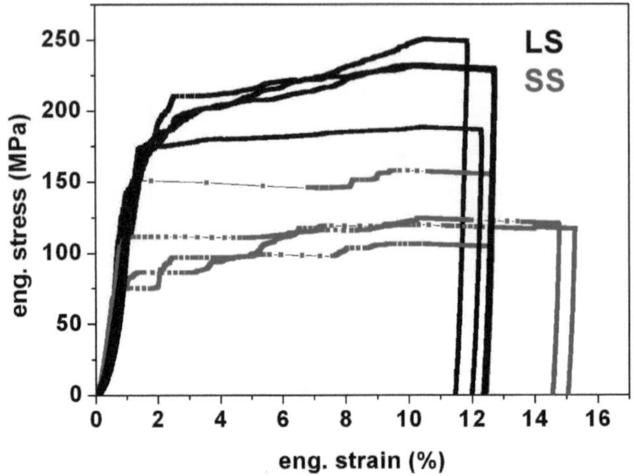

Figure 5.4: Stress strain curves obtained from cuboidal Cu pillars having geometry 1.

Figure 5.5: SEM micrographs: Deformed Cu pillars : (left) In the LS configuration slip lines (parallel to the red dotted line) run into the substrate (right) In the SS configuration the pillar deforms without influence of the substrate.

Geometry 2

For this geometry (5:3:1 μm^3, H:L:W) the Cu pillars exhibit no difference in flow stress for the LS and SS orientation (figure (5.6)). The stress-strain curves show a slightly different strain hardening behaviour. Strain hardening in the SS configuration occurs in a stair-like manner whereas it is smoother in the LS case. This effect can be explained by the deformation behaviour of the two different orientations. In the SS configuration the pillar top shears off and the indenter moves abruptly until it reaches the remaining part (indicated by a red arrow in figure (5.7)) of the pillar which leads to the observed steps. In the LS configuration additional glide systems have been activated (indicated by a blue arrow in figure (5.7)) which can be attributed to stress concentrations that arise because of the friction between the indenter tip and the pillar top.

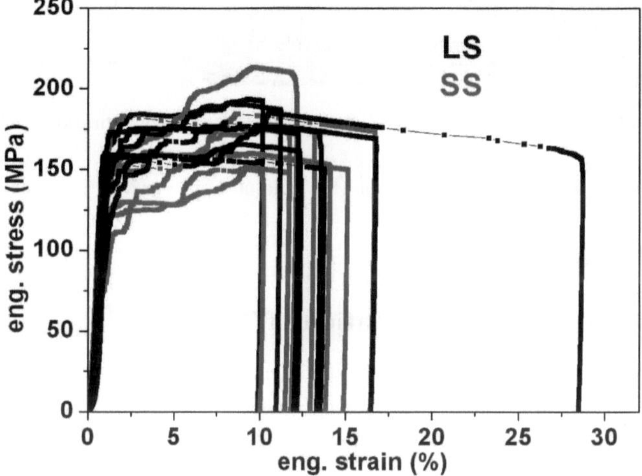

Figure 5.6: Stress strain curves obtained from cuboidal Cu pillars having geometry 2.

Figure 5.7: SEM micrographs: Compressed Cu pillars: (left) LS configuration (right) SS configuration
Arrows indicate activation of second slip system (left) and unslipped part of the pillar (right)

5.2 Tantalum

5.2.1 Size Dependent Deformation of Ta

Figure (5.8) shows typical examples of stress-strain curves for different column diameters. The stress strain curves show that higher stresses are needed in order to deform smaller pillars. It can be seen that the 7.84µm and 1.27µm columns deformed steadily up to about 10% whereas all other columns deformed abruptly by a more or less pronounced displacement burst exceeding the 10% strain limit. A general correlation between the magnitude of the strain bursts or the resulting deformation velocity with the column diameter could not be established from the data.

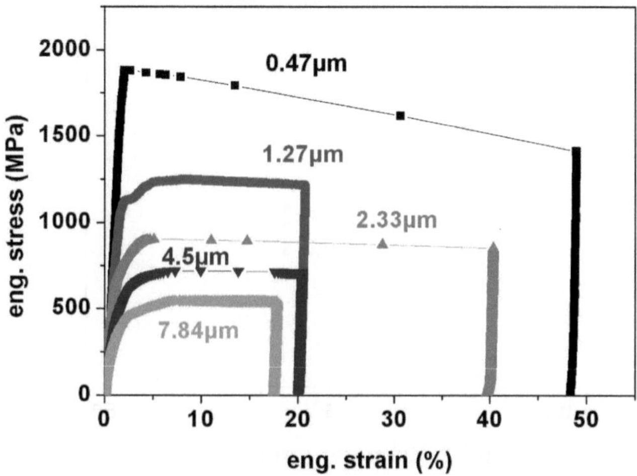

Figure 5.8: Representative stress-strain curves for ⟨111⟩ oriented Tantalum micro-pillars. The numbers next to the curves indicate the diameters of the pillars.

Figure (5.9) summarizes the flow stresses at 5% plastic strain for ⟨111⟩- ,⟨100⟩- and ⟨235⟩-oriented micro-pillars as a function of pillar diameter. The ⟨111⟩- and the ⟨100⟩-pillars were milled into individual grains of the same poly-crystalline Ta sheet. The orientations of the grains were determined by EBSD and a misorientation of up to 10° was tolerated. The pillars with the ⟨235⟩-orientation were machined into a single crystal.
The solid lines in figure (5.9) are a linear regression through the data. The slopes of the regression lines represent the power law exponents according to equaton (2.16). Furthermore, it can be seen that the scatter of the data increases as the diameters of the columns become smaller.

The SEM examination of the tested ⟨111⟩-columns revealed different deformation morphologies. Some of the columns deformed by localized glide only on one active glide plane (figure (5.10(a))). Others deformed by glide on different glide planes (figure (5.10(b))) and sometimes a barrel shaped distortion without detectable slip steps was observed (figure (5.10(c))). From 36 investigated columns 16 columns showed localized slip, in 14 columns glide on multiple slip planes was found and in 6 columns a barrel

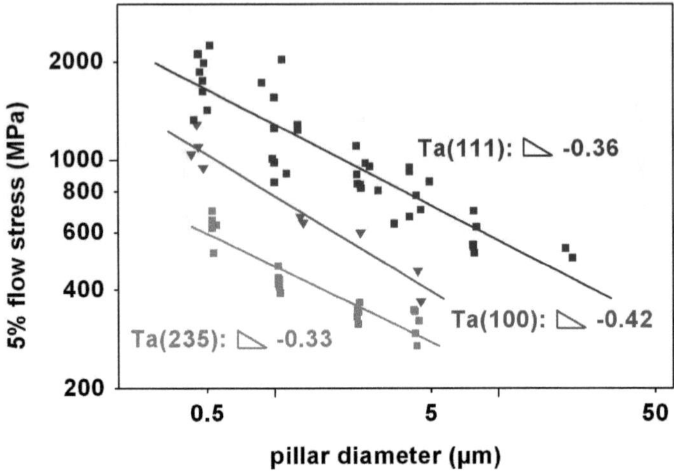

Figure 5.9: Flow stress at 5% plastic strain as a function of pillar diameter. The solid lines are a linear regression through the data points. The slope of these lines corresponds to the size effect exponent β.

shaped distortion was observed. From the data a correlation between column diameter and deformation morphology could not be identified. A rather large misorientation of 10° from the $\langle 111 \rangle$ direction was tolerated. This leads to different activated glide systems and may explain the different observed deformation morphologies.

The pillars machined in the single-slip oriented crystal deform in the same manner with different activated glide systems at the pillar top and only one activated glide system distant from the pillar top (figure (5.11)). The complex deformation morphology at the pillar top can be attributed to friction stresses between pillar and indenter tip.

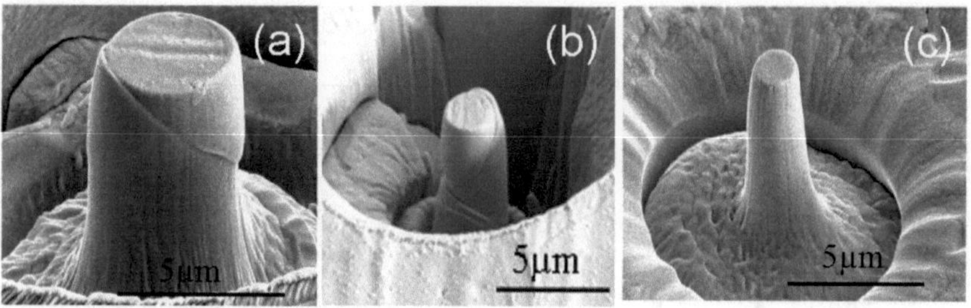

Figure 5.10: SEM micrographs of deformed pillar showing different deformation morphologies: (a) The column deforms by localized glide on one glide plane. (b) Two or more glide systems are active. (c) A barrel-shaped distortion.

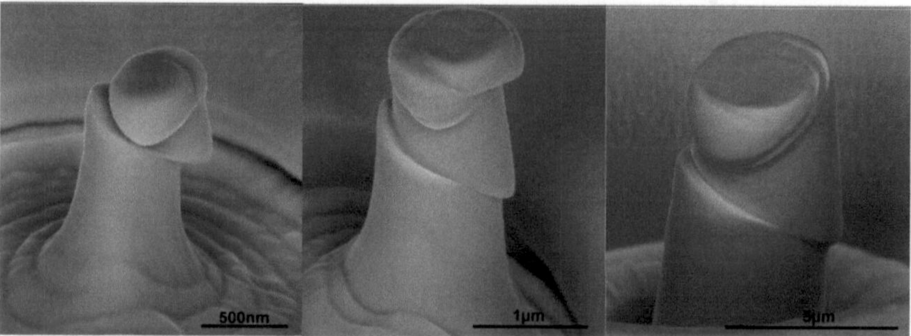

Figure 5.11: SEM micrographs: Deformation morphology of ⟨20 7 45⟩ oriented Ta pillars

5.2.2 Load Rate Dependence

⟨111⟩ oriented Ta pillars with a diameter of 4μm were compressed with varying load rates. The load rates were in a range between $0.03125\frac{\text{mN}}{\text{s}}$ and $5\frac{\text{mN}}{\text{s}}$. The measurements were made at room temperature, well below the athermal temperature of Ta (450K [86]). The results of these measurements are shown in figure (5.12). It can be seen that the flow stress depends on the load rate at which the compression experiment is performed. A detailed discussion about this phenomenon is given in chapter (6.9).

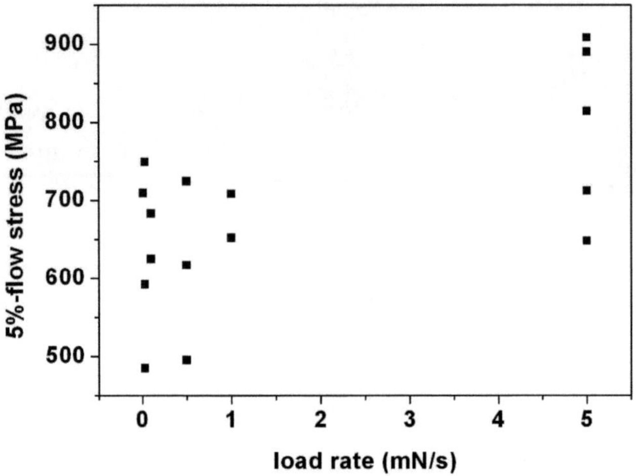

Figure 5.12: Flow stress of 4μm Ta-pillars with respect to load rate. The load was applied along the ⟨111⟩ direction.

5.2.3 Deformation of Cuboidal Ta Pillars

Geometry 1

Ta shows significantly higher stresses for the LS configuration than for the SS configuration (figure (5.13)). Investigations of the compressed pillars (figure (5.14)) show that in this case it is not the substrate effect (cf. chapter (5.1.2)) that was observed in Cu. To ensure that the higher stresses do not originate from the substrate, Ta was also tested with geometry 2 (figure (5.15)).

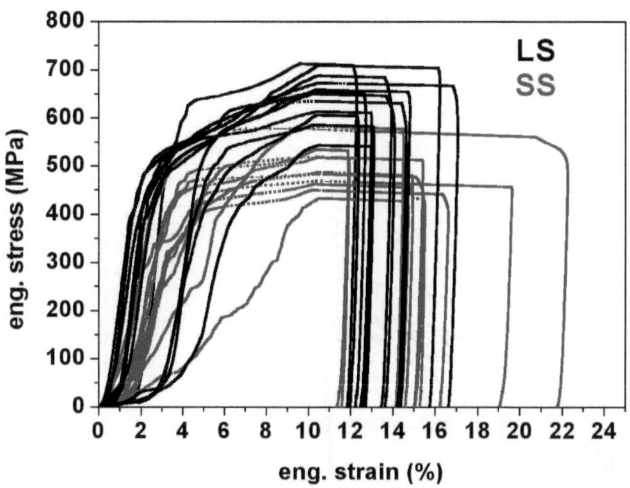

Figure 5.13: Stress strain curves obtained from cuboidal Ta pillars having geometry 1.

Figure 5.14: SEM micrographs: Compressed Ta pillars (geometry 1): (left) LS configuration (right) SS configuration

Geometry 2

As already observed for geometry 1 the short cuboidal Ta pillars show an orientation dependent stress-strain behaviour (cf. figure (5.15)). Stresses in the LS configuration are higher than in the SS configuration. Contrary to Cu, the effect is observable for both geometries. This strongly suggests that the effect of the higher stresses for the LS configuration is a real effect of the material and does not originate from the geometry of the pillars. Representative compressed pillars are shown in figure (5.16).

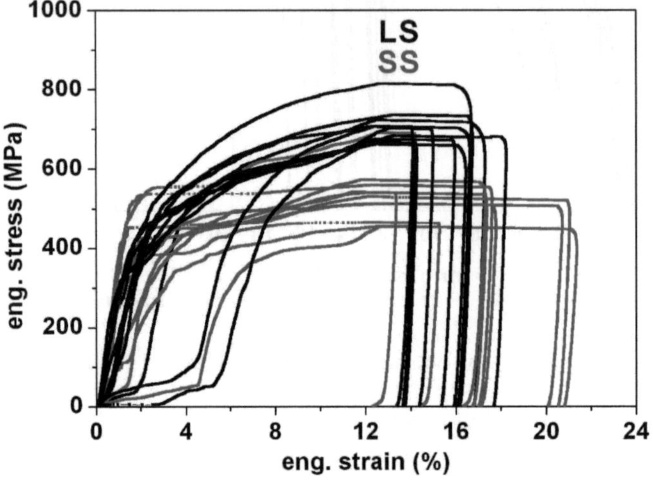

Figure 5.15: Stress strain curves obtained from cuboidal Ta pillars having geometry 2.

Figure 5.16: SEM micrographs: Compressed Ta pillars (geometry 2): (Left) LS configuration (Right) SS configuration

5.2.4 EBSD After Compression

EBSD measurements are used in order to investigate the deformation behaviour of microcompression experiments. This can be done by mapping the lateral surface of a compressed pillar. An image of the pillar is shown in figure (5.17). The pillar needs to be milled at the edge of the sample to avoid shadowing of the backscattered electrons. When doing an EBSD measurement on the sample surface the sample is usually tilted to 70°. For an EBSD scan of the pillars lateral surface the sample needs to be tilted to −20°. The result of the EBSD measurement of the compressed pillar is shown in figure (5.18) which is a map of misorientations with respect to a defined orientation. This defined orientation was taken from the base of the pillar where little deformation can be expected. It can be seen that the pillar exhibits rather large misorientations of up to 10°. The highest values of misorientations are found along the slip planes and at the pillar top.

Figure 5.17: SEM micrographs: Compressed Ta pillar. The images are rotated by an angle of 90° against each other.

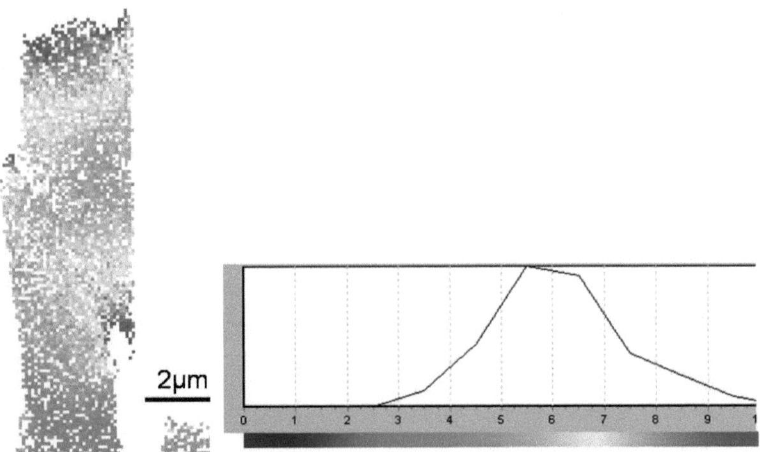

Figure 5.18: EBSD after compression of a Ta pillar. The legend shows the misorientation with respect to the substrate in degrees.

5.2.5 In Situ Microcompression

With a special nanoindenter (manufactured by Hysitron, Minneapolis, USA) it is also possible to perform compression tests inside the SEM chamber. The advantage of this technique is that deformation behaviour of the pillar can be observed in real time. It is possible to see whether deformation processes coincide with the occurance of load drops in the force displacement response. The procedure to perform such a test is rather complex and therefore it was only performed once. Figure (5.19) shows the compression of a Ta pillar inside the SEM. The image shows two pronounced slip lines where the slip line at the pillar top was observed first. By taking a closer look, it can be seen that there are slip lines distributed over the entire pillar. These slip lines show that not only one glide plane was activated during compression. Although the pillar was machined by the lathe cut technique there seem to be stress concentrations at the pillar top which are most likely associated with friction between the indenter and the pillar.

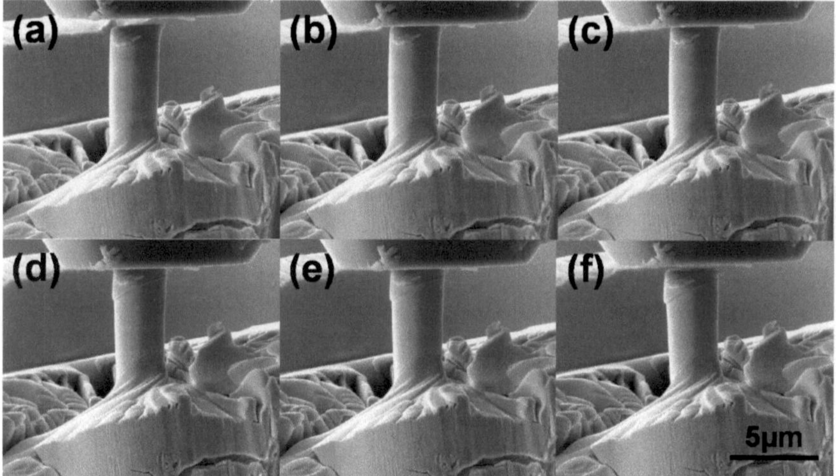

Figure 5.19: SEM micrographs: In-situ compression of a Ta pillar. (a) indenter is moving towards the pillar, but is not yet in contact, (b) and (c) indenter is in contact and continues its downward movement, (d) slip line forms at pillar top, (e) and (f) indenter is unloading.

Figure (5.20) shows the load-displacement response of the pillar from Figure (5.19). The pillar was loaded up to a load of 3000µN which corresponds to ≈ 50MPa. It can be seen that the indenter did not touch the pillar until a displacement of 300nm was reached.

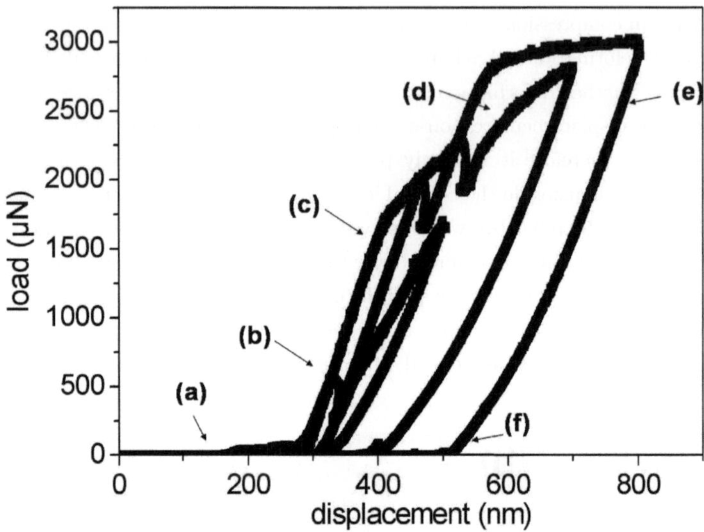

Figure 5.20: Load-displacement curve of an in-situ compressed Ta-pillar. The maximum load of 3000µN corresponds to ≈ 50MPa. (a)-(f) correspond to the images in figure (5.19).

5.2.6 Investigation of the Effect of Initial Dislocation Density on Pillar Deformation

The initial dislocation density affects the deformation behaviour of samples with small volumes. Recent studies suggest that a size effect cannot be observed in the absence of initial dislocations [46, 59, 74, 87]. The influence of the intial dislocation density can be investigated by milling pillars into regions with a low dislocation density and regions with a high dislocation density. Some deep Berkovich indents (indentation depth=2µm) were performed in order to increase the number of dislocations locally. Pillars with a diameter of ≈ 3µm and a height of ≈ 12µm were machined in the vicinity of the indent (cf. figure (5.21)).
The stress strain curves of the pillars reveal that there is not a significant difference between pre-damaged samples and standard samples (figure (5.22)). This could mean that the initial dislocation density is already above a certain level and that the investigated samples contain enough dislocations so that a further increase does not have a

significant effect on the deformation behaviour. This may, however, change when the sample size is decreased.

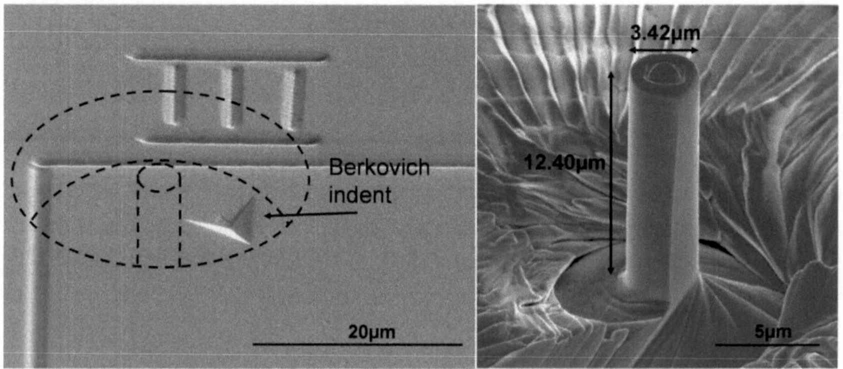

Figure 5.21: SEM micrographs: Fabrication of a pre-damaged sample.

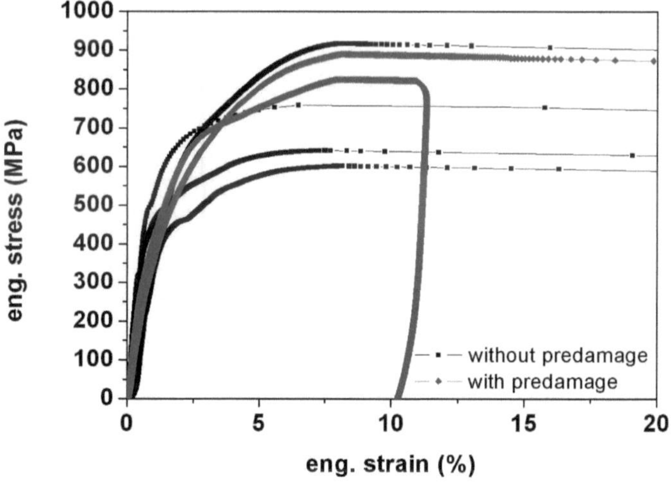

Figure 5.22: Stress strain curves for undamaged samples (black) and pre-damaged samples (red).

5.3 alpha-Iron

5.3.1 Size Depependent Behaviour

The microcompression experiments were performed on the sample described in chapter (3.7.3). Fe shows a strong size dependent behaviour with smaller samples exhibiting higher flow stresses. The stress strain curves of these experiments are similar to the ones observed for Ta and Cu and do not show a qualitative difference. The large pillars show a bulk-like deformation behaviour whereas the small pillars show stochastic behaviour with pronounced strain bursts (cf. figure (5.23)). All tested pillars were milled into

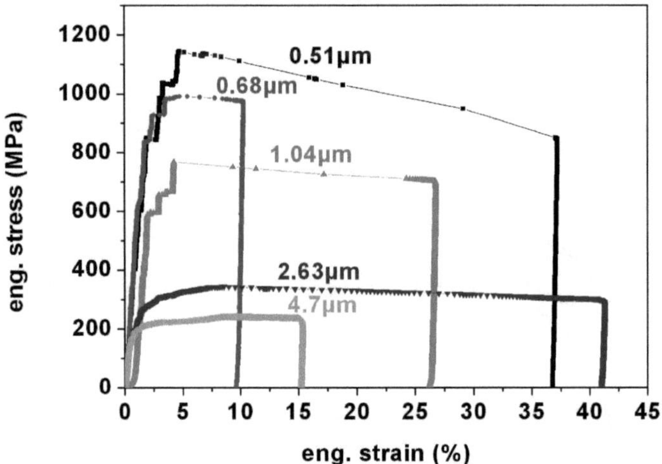

Figure 5.23: Representative stress strain curves for microcompression experiments made on cylindrical Fe pillars. The numbers next to the curves give the pillar diameter.

the same grain, and the pillars show the same deformation morphology for all pillars, as can be seen in figure (5.24).

Figure (5.25) shows a double logarithmic plot of the 5%-flow stresses with respect to pillar diameter. The power law exponent β was found to be -0.81.

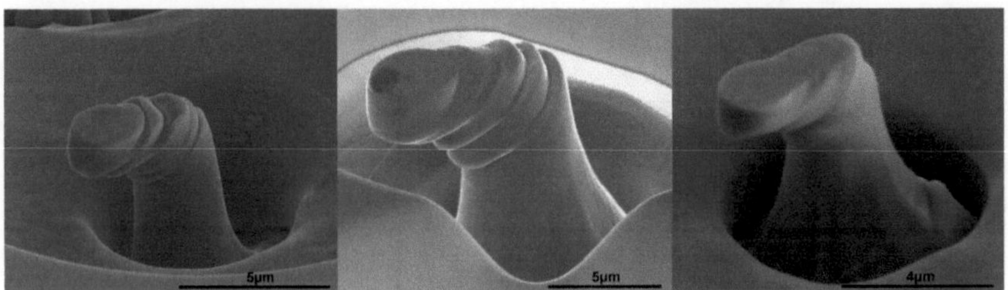

Figure 5.24: SEM micrographs: Representative compressed Fe pillars showing a uniform deformation morpholoy.

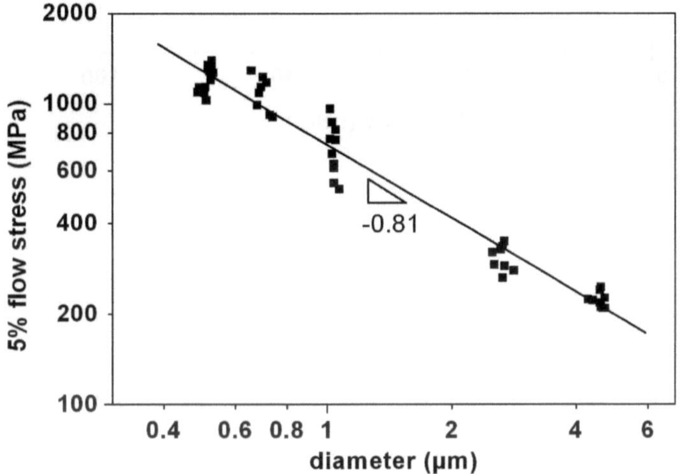

Figure 5.25: Size effect in Fe. Stress values are taken at 5% plastic strain.

5.3.2 Load Rate Dependence

The load rate dependence of Fe was measured by means of microcompression and nanoindentation. Microcompression experiments were performed on pillars with diameters of 0.5μm (figure (5.26)) and 1μm (figure (5.27)). The out-of-plane orientation of the tested sample was close to $\langle 211 \rangle$. The loading rates of the 0.5μm pillars were varied between $0.3\frac{\mu N}{s}$ and $70\frac{\mu N}{s}$. The variation of the load rates for the 1μm pillars was in a range between $1\frac{\mu N}{s}$ and $300\frac{\mu N}{s}$.

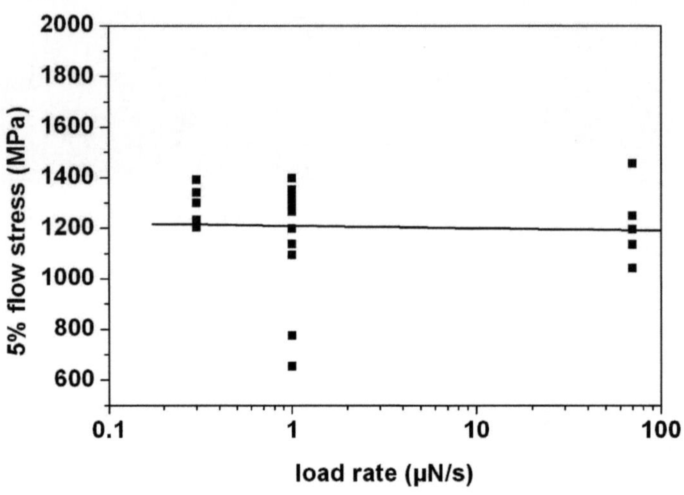

Figure 5.26: Load rate sensitivity of 0.5µm Fe pillars.

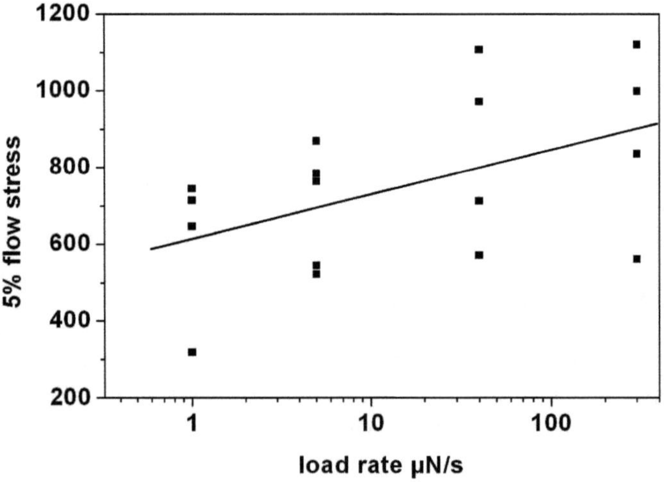

Figure 5.27: Load rate sensitivity of 1µm Fe pillars.

Indentation Strain Rate Dependence

The nanoindentation experiments were performed with a Berkovich indenter into the same grain as the experiments in chapter (5.3.2). Three different indentation strain rates were applied, namely $1.5 \cdot 10^{-1}\mathrm{s}^{-1}$, $1.5 \cdot 10^{-2}\mathrm{s}^{-1}$ and $1.5 \cdot 10^{-3}\mathrm{s}^{-1}$ (a defintion of the indentation strain rate is given in the Appendix (A.2). For each strain rate, 15 indents were performed. Figure (5.28) shows three individual indentation test with pronounced pop-in behaviour, i.e. the indenter penetrates into the material without increasing force. A pop-in is considered to occur when dislocations form underneath the indenter and can be considered as transition from elastic to plastic behaviour in a nanoindentation experiment [88]. For an ideal Berkovich indenter with a perfectly sharp tip, pop-ins should not be observed. The observed pop-in behaviour can therefore be considered as indication that the indenter tip is blunted. A tip area calculation was performed in order to take the non-ideal tip geometry into account.
For an indentation experiment with a spherical indenter the elastic part of the force-displacement curve can be fitted by using the 'Hertz'-equation [89]:

$$F = \frac{4}{3}E^{*}\sqrt{R}h^{\frac{3}{2}} \tag{5.1}$$

with load F, reduced modulus E^{*}, tip radius R and indentation depth h.
The reduced modulus E^{*} represents a composite measure of the compliance of the sample and of the indenter:

$$\frac{1}{E^{*}} = \frac{1-{\nu_i}^2}{E_i} + \frac{1-{\nu_s}^2}{E_s} \tag{5.2}$$

With the values of diamond ($E_i = 800\mathrm{GPa}$, $\nu_i = 0.07$ [82]) and Fe ($E_s = 211\mathrm{GPa}$, $\nu_s = 0.293$ [90]) the reduced modulus is determined to $E^{*} = 179\mathrm{GPa}$. In this case the 'Hertz'-equation can be used to estimate the tip radius of the Berkovich indenter which is a measure of the sharpness of the tip. With values from the nanoindentation experiment and $E^{*} = 179\mathrm{GPa}$ a tip radius of $\approx 85\mathrm{nm}$ is calculated.
Figure (5.29) shows the mean load displacement curves of each individual strain-rate. The dashed line represents the Hertzian fit of the elastic part of the indentation curves. It can be clearly seen that the strain-rate affects the load-displacement behaviour of the indentation experiments.

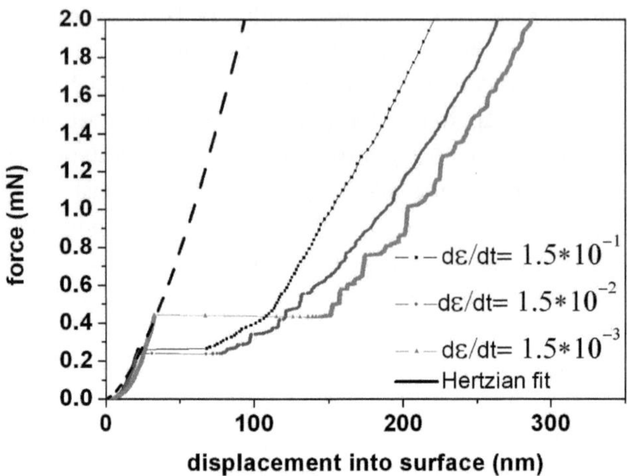

Figure 5.28: Three individual load-displacement curves for nanoindentation in Fe. The different curves show pronounced pop-in events.

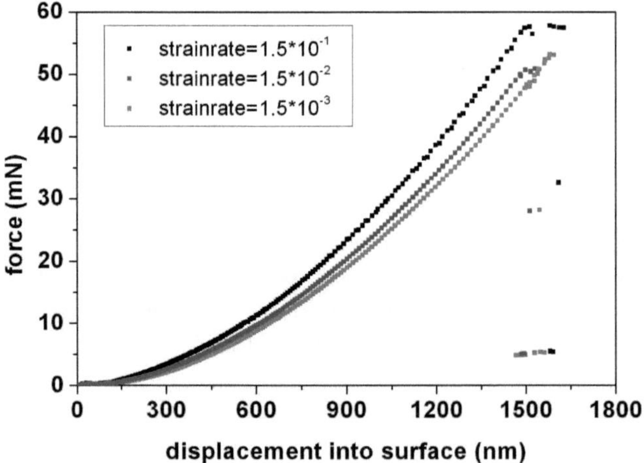

Figure 5.29: Indentation experiments on Fe exhibit strain rate dependency. Each curve represents the mean force displacement behaviour of 15 individual measurements.

5.3.3 Deformation of Cuboidal Fe Pillars

The cuboidal Fe pillars do not show a difference in stress for the two different orientations. But there is a difference in strain although all pillars were intended to be compressed to approx. 10% of their height. The higher strains for the SS configurations were observed for almost all microcompression experiments. An explanation of this behaviour can be found in chapter (6.6). The compressed pillars did not show well resolved slip lines (figure (5.31)) which can be attributed to cross slip and multiple active slip systems. The cuboidally shaped Fe-pillars were only tested in geometry 1. It can be argued that the material at the base affects the stress strain behaviour as well, as it is observed for Cu. There are two explanations why this strengthening of the LS orientation does not occur in Fe. First, the angle between the activated glide system and the sample surface may be small enough that the slip lines do not run into the base of the pillar. Second, if the slip lines run into the base material and the stress increases, another glide system will be immediately activated. This glide system requires a higher stress, but due to the amount of 48 possible glide systems, it is very likely that only a small stress increase is necessary to activate this glide system.

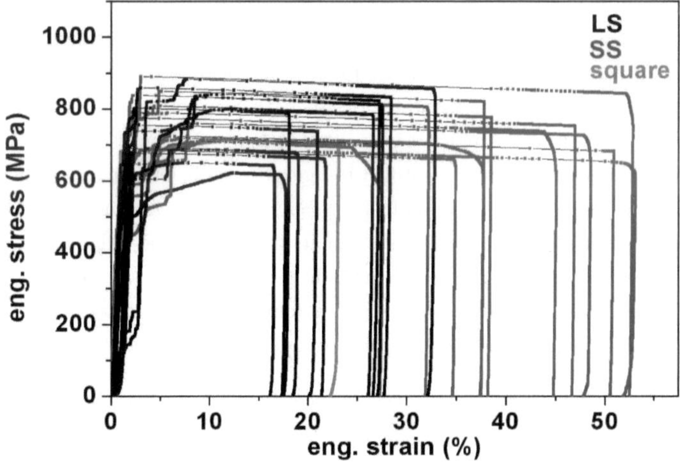

Figure 5.30: Stress strain curves obtained from cuboidal Fe pillars having geometry 1.

Figure 5.31: SEM micrographs: Compressed Fe pillars: (left) LS configuration (right) SS configuration

5.4 Identification of Activated Glide Systems

In this work a method was developed which allows the identification of glide planes. The technique is explained in detail in chapter (3.6). In order to confirm whether the technique works, it was applied to Cu, since the fcc lattice has only twelve possible glide systems which makes it easier to identify them. One example of the glide plane identification procedure is shown in figure (5.32). The Schmid factors of the possible glide systems have been calculated and the results have been compared to the identified glide planes. The highest Schmid factor m=0.44 is found for the $[011]\left(1\bar{1}1\right)$ system. The (111) glide plane which was identified in figure (5.32) belongs most likely to the $\left[0\bar{1}1\right](111)$ glide system with a Schmid factor m=0.31. Considering the Schmid factors this glide system should not be activated, but its location at the pillar top indicates that the friction between the pillar and the indenter leads to stress concentrations so that these systems can be activated. Six pillars have been investigated with 14 glide planes of the $\left(1\bar{1}1\right)$ type. The mean normal vector $\vec{n}$ of these glide planes has been determined and looks as follows,

$$\vec{n} = \begin{pmatrix} 0.6597 \pm 0.03873 \\ -0.4860 \pm 0.02632 \\ 0.5700 \pm 0.04238 \end{pmatrix} \tag{5.3}$$

with a mean misorientation from the $\left[1\bar{1}1\right]$-direction of $7.06° \pm 2.5°$. For comparison, the misorientation from the (111) plane is 64.53°, from the $\left(\bar{1}11\right)$ plane 70.55° and from the $\left(11\bar{1}\right)$ plane 76.74°. There is no doubt that the identified glide plane is the

$(1\bar{1}1)$ plane. From the small standard deviation values in $\vec{n}$ can be seen that the error in orientation occurs systematically. The misorientation of $\approx 7°$ can be understood by considering the consecutive steps of the glide plane measurement. The sample has to be taken out of the SEM after EBSD in order to do the compression experiments. After the experiments the sample has to be mounted in the SEM again for imaging. This procedure may produce a small amount of misalignment. Furthermore the compression of the pillars may bend the pillars or lead to crystal rotation. As can be seen in chapter (5.2.4) misorientations of $\approx 5°$ are not unlikely to occur. The identification of glide

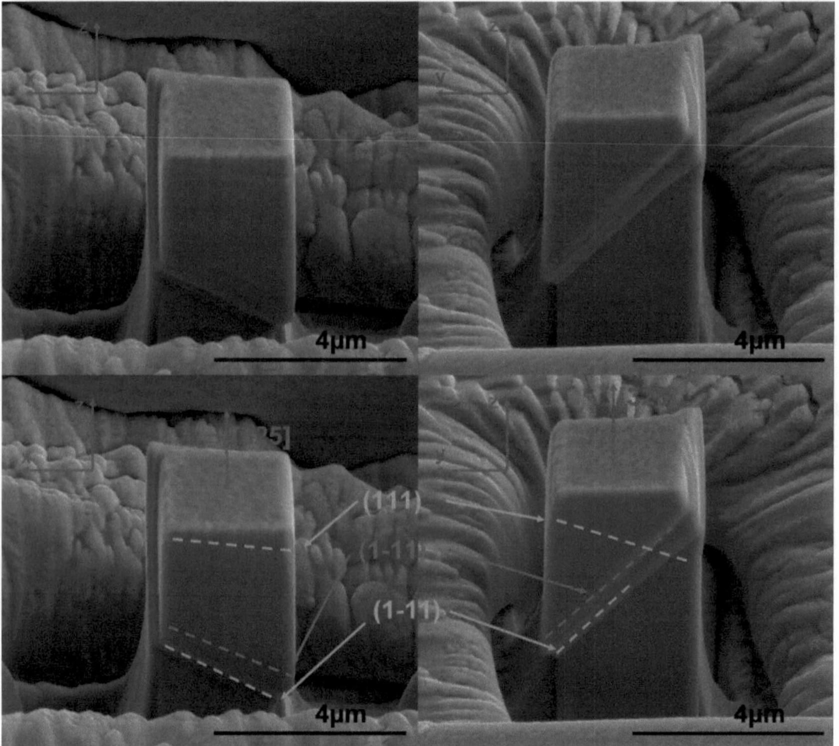

Figure 5.32: SEM micrographs: Identification of glide planes in Cu. The calculated glide planes have misorientations between 4° and 6° from the glide planes indicated in the figure.

planes becomes more complex when metals with a bcc lattice structure are considered. Instead of 12 different glide systems 48 systems are available, if the {321} glide planes are taken into account. Additionally cross slip processes, which are more likely to occur in bcc metals than in fcc metals complicate glide plane identification. Both techniques, identification on pillars with square cross section and cylindrical pillars (cf. chapter (3.6)), have been applied to Ta. The technique for cylindrical pillars was only applied once, since fitting the geometrical shapes on the SEM image is always subject to errors. Glide plane identification using pillars with square cross section is therefore supposed to be more accurate. An image of the cylindrical pillar is shown in chapter (3.6). The out-of-plane orientation of the pillar is close to [111], for a more accurate determination of the glide plane Euler angles obtained from an EBSD measurement were used. The investigation of the pillar resulted in the following normal vector $\vec{n}$ of the glide plane

$$\vec{n} = \begin{pmatrix} 0.1 \\ 0.67 \\ 0.73 \end{pmatrix} \tag{5.4}$$

with a misorientation of $\approx 13°$ from $\{321\}$, $\approx 24°$ from $\{211\}$ and $\approx 6°$ from $\{110\}$. From this it seems that glide is most likely to occur on the (011) plane. A consideration of the Schmid factors shows that the $[1\bar{1}1]$ (011) system has a relatively high Schmid factor of m=0.290458, but it is not the highest Schmid factor that can be found. The highest Schmid factor m=0.340949 is calculated for the $[1\bar{1}1]$ (132) system. BCC metals do not necessarily obey the Schmid law, the packing density of the glide plane and the mobility of the dislocations on specific planes have to be taken into account. Furthermore slip on $\{321\}$ is rarely observed. The result is reasonable and confirms that the technique for cylindrical pillars works, although it is not as convenient and precise as the glide plane identification of the pillars with the square cross section.

The glide plane identification was also applied on Ta-pillars having a square cross section. Four pillars having an out-of-plane orientation close to [325] were tested. An example is shown in figure (5.33). The slip lines are not as clear as in the case of Cu. The pillars showed wavy slip lines which is a consequence of the ease of cross slip in bcc metals. All investigated pillars exhibited the (011) plane. The misorientations between the observed glide planes and the (011) plane were in a range between 1.6° and 7.8° which is even smaller than the misorientations observed in Cu. In the particular case of the pillar in figure (5.33) a second glide plane could be identified. This second glide plane shows a misorientation of 7.96° from the (121) and 3.03° from the (132) plane, so glide on the (132) plane seems to be more likely. The Schmid factors of all 48 possible glide systems were calculated and compared to the glide planes that have been found. Table (5.1) shows the glide systems with the 5 highest resolved shear stresses.

The Schmid factors and the analysis of the SEM images indicate that glide takes place on the $[1\bar{1}1]$ (011) glide system and that cross slip to the $[1\bar{1}1]$ (132) system or the $[1\bar{1}1]$ (121) system is possible.

Glide System	Schmid factor
$[1\bar{1}1]$ (132)	0.461932
$[1\bar{1}1]$ (011)	0.442161
$[1\bar{1}1]$ (121)	0.424664
$[1\bar{1}1]$ ($\bar{1}$23)	0.405217
$[1\bar{1}1]$ (231)	0.403575

Table 5.1: Glide systems with the highest Schmid factors for [325] oriented Ta.

It was also attempted to apply the technique on α-Fe pillars but the slip lines in these samples were not clear enough. This may be caused by the high mobility of the screw dislocations. The screw dislocations may change their glide planes very often, so that slip occurs in a wavy manner and therefore no clear slip lines can be observed.

Figure 5.33: SEM micrographs of a representative Ta pillar. The images in the first row are rotated by 90°around the pillar axis against each other. The images in the second row show the identified glide planes. The projection of the Burgers vector $\vec{b}$ onto the sample surface is indicated by a red arrow.

6

Discussion

In this chapter the size dependent mechanical behaviour of Ta and Fe is discussed and compared to other studies from the literature. Different normalization procedures will be presented in order to investigate whether size effects in fcc and bcc metals originate from the same underlying mechanisms. A comparison of different bcc metals shows that the size effects seem to be temperature dependent which may be attributed to the low mobility of screw dislocations. Different possible mechanisms are described that may explain the observed small-scale plasticity of bcc metals. The role of screw dislocations is investigated and their interaction with free surfaces. The differences of small-scale plasticity of fcc and bcc metals is described by a composite material model which is based upon the assumption that surfaces enhance the mobility of screw dislocations in bcc metals. At the end of this chapter the strain rate sensitivity and the activation volume of small Ta samples is discussed.

6.1 Mechanical Size Effects in Ta

Figure (5.9) shows the flow stresses of Ta for different column diameters and different orientations. Flow stresses of columns with $\langle 111 \rangle$ orientations are higher than the ones with $\langle 100 \rangle$ orientation. Although dislocations in bcc metals do not always glide on atomic planes with the highest resolved shear stress [24](cf. chapter (2.3)), the Schmid factors of both orientations are compared. The Schmid factors of the glide systems containing $\{110\}$ planes in the $\langle 111 \rangle$ oriented columns are smaller by a factor of ≈ 1.5 than those of the $\langle 100 \rangle$ orientation which is consistent with the difference in the flow stress. The pillars with the $\langle 235 \rangle$ orientation are somewhat weaker than the pillars for the two other orientations. This can be expected since the activated glide system has a high Schmid factor (≈ 0.5) and strain hardening due to the interaction of different activated glide systems is unlikely to occur.

The dependence of the flow stress on the column diameter is in the same range for all orientations. The lines of best fit according to equation (2.16) are less steep than the ones found for fcc metals (table (6.2)). Studies on Mo [10, 80] and Nb [79] confirm the trend of higher absolute stresses and less pronounced size dependencies of bcc metals compared to fcc metals. Extrapolating the line of best fit towards large sample size reveals the size at which the mechanical behaviour of the pillars coincides with that of the bulk material. Uchic et al. [6] reported that Ni pillars of a diameter of 20µm show bulk-like behaviour which is in good agreement with the Cu data shown in figure (5.3). For Ta, samples with diameters of 20µm have only been tested for the ⟨111⟩ orientation (figure (5.9)) and they still exhibit stresses that are well above the bulk yield strength, which is approximately 150MPa [13] for the strain rate used in these experiments. It is therefore likely that the transition to bulk-like behavior occurs for larger sample sizes on the order of 100µm in diameter.

6.2 Mechanical Size Effects in Fe

Size-dependent behaviour is also observed when the microcompression technique is applied to Fe pillars. As for the other materials studied in this work, strengthening is observed when the pillars are decreased in size. The stress-strain curves show a behaviour as it is observed for many fcc metals with bulk-like deformation for larger pillars and stair-like behaviour for smaller pillars (cf. figure (5.23)). The reason for this behaviour can be explained by single dislocation activities which are more dominant in smaller pillars than in larger ones. Deformation in larger pillars occurs by the motion and interaction of dislocation networks, whereas in smaller pillars only a small number of dislocation sources is active and the activation of a single dislocation source might lead to a strain burst in the stress-strain response.

Figure (5.25) shows the flow stresses at 5% plastic strain with respect to the pillar diameter. According to equation (2.16) the slope of the linear fit corresponds to the power law exponent. The power law exponent found here is -0.81 and is approximately twice as large as the exponents found for Ta. A value of -0.81 is comparable to the values observed for fcc metals. The increase in stress with decreasing pillar size is therefore much more pronounced in Fe than in Ta. The stronger size effect in Fe compared to Ta may be attributed to the different athermal temperatures of both metals. The athermal temperature of Ta is higher than the one of Fe which means that screw dislocation motion in Ta requires more thermal activation than screw dislocation motion in Fe. If the athermal temperature of a bcc metal is below room temperature (all compression tests were performed at room temperature) an fcc-like deformation behaviour of the bcc metal is not surprising. A more detailed discussion about the power law exponents and thermally activated screw dislocation motion in bcc metals is given in chapter (6.4).

6.3 Comparison of the Mechanical Size Effect of fcc and bcc Metals

In order to compare the size dependence of the yield strength of different materials, the data need to be normalized. Table (6.2) gives an overview of the power law exponents found for different materials. For fcc materials, the flow stresses are commonly normalized with respect to the shear modulus μ, and the Burgers vector b. The value for $\mu \cdot b$ is proportional to the theoretical strength of the metal or the strength of a Frank-Read source of a given size [91]. It has been shown that this normalization procedure works well for fcc metals where the bulk yield strengths are relatively small compared to the stresses arising from the size effect [7]. Normalization becomes less straightforward when materials of different isomechanical groups [92] are to be compared. When comparing bcc with fcc materials, a normalization as described above may not be appropriate because the normalized values for the bulk strength of the different materials do not coincide. For fcc metals, σ_0 in equation (2.16) is significantly smaller than $c \cdot d^{-\beta}$ and can be neglected for the sample size considered here. This is not the case for bcc metals where the bulk stresses are often in the range of 100MPa and higher. Therefore another approach for normalization has to be considered. The bulk yield strength is subtracted from the measured flow stress and then the result is normalized by $\mu \cdot b$. This treatment of the data leads to a different power low exponent of the metals. Figure (6.1) shows the normalized data and it can be seen that even after applying this procedure, Ta still shows higher flow stresses and has a less pronounced size effect compared to Cu and Al.

Instead of subtracting the bulk yield strength, the curves can be scaled by a factor so that the bulk stresses of the individual metals are equalized. When this procedure is applied to the Ta data, the stress level of Ta approaches those of fcc metals but the smaller slope still persists. Different normalization procedures have been applied, but no normalization procedure was found that can account for the less pronounced size dependence that is found for Ta. It is therefore justified to assume that there are fundamental differences in the size dependent behaviour of fcc and bcc metals.

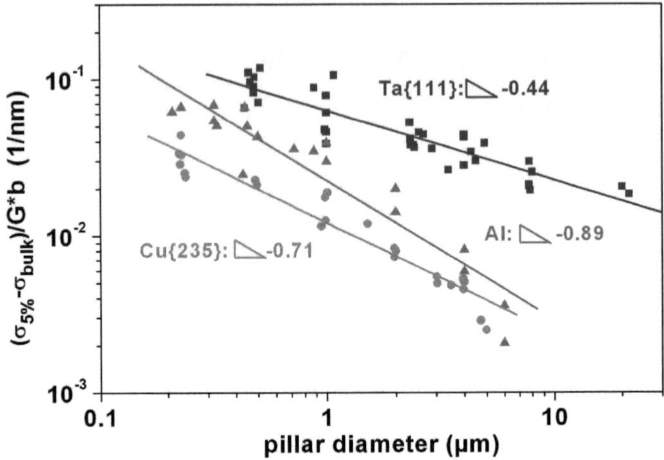

Figure 6.1: Size effect in Ta, Cu and Al [8]: Flow stresses are plotted versus pillar diameter after subtracting the bulk yield strength and normalizing the data by $\mu \cdot b$.

6.4 Comparison of Different bcc Metals

As has been shown in chapter (6.3) there are fundamental differences in the mechanical size effects of bcc and fcc metals. The difference between the power law exponents of α-Fe and Ta is quite large and it seems that there is no common scaling behaviour in bcc metals as it is observed for fcc metals (cf. figure (2.11) and table (6.2)).

A comparative study of four different bcc metals (W, Mo, Ta and Nb) gives new insights into the nature of the size effect [79]. The pillars were $\langle 001 \rangle$ oriented and the tests were performed at room temperature. The different bcc metals exhibit different athermal temperatures T_{ath} (table 6.1). Above the critical temperature the flow stress of the metal is insensitive to the test temperature which means that enough thermal energy is available that the screw dislocations can overcome the Peierls barrier. Therefore screw and edge dislocations have comparable mobilities above T_{ath}.

Material	Fe	Nb	Ta	Mo	W
athermal temperature T_{ath}(K)	300	350	450	480	800
Temperature ratio $\frac{T_{test}}{T_{ath}}$	0.99	0.85	0.66	0.62	0.37

Table 6.1: T_{ath} for different bcc metals according to [86, 93–99]

The stress strain curves of the different bcc metals show the commonly observed behaviour with bulk-like deformation for large pillars and a disruptive deformation behaviour for pillars approximately smaller than 2µm in diameter. The deformation morphology of the pillars changes significantly with size. The larger pillars with a diameter of 2µm or larger show a deformation morphology which seems to depend on the athermal temperature of the material. In Nb and Ta (low T_{ath}) localized slip is observed whereas Mo and W (high T_{ath}) deform by wavy slip. Generally speaking deformation in bcc metals is controlled by the motion of long and straight screw dislocations [28]. This indicates that more cross-slip takes place in metals with high T_{ath}. For metals with low T_{ath} the pure screw dislocations may bow out and deviate from the pure screw character leading to less cross-slip events since the mixed dislocation is confined to a specific glide plane. For small pillars no influence of the athermal temperature on the deformation morphology can be observed, these pillars deform by localized slip. The periodicity of wavy slip might be on the order of the pillar diameter and are therefore not observable at this size scale. Another explanation could be that deformation in the smallest pillars is controlled by the nucleation of dislocations rather than by their interaction. A Frank-Read source emits mixed dislocations and these dislocations might encounter the surface before they can straighten up to pure screws leading to less cross-slip.

All metals investigated show a pronounced mechanical size effect where a decrease in dimensions leads to a strenghtening of the material. As already observed for Ta and Fe, the scaling behaviour of the different metals differs significantly. The flow stresses at 5% strain with respect to pillar diameter are shown in figure (6.2).

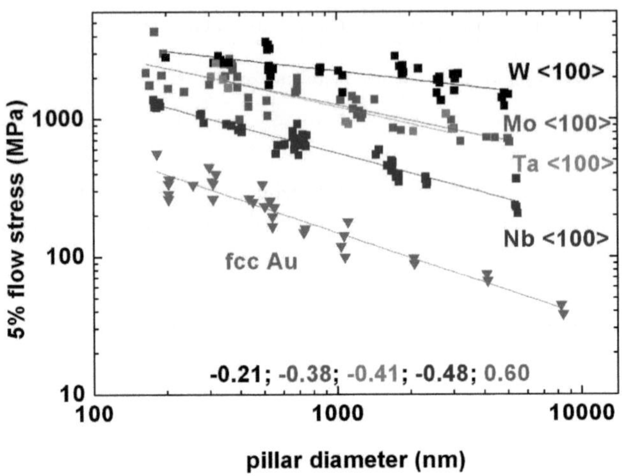

Figure 6.2: Size effects of four different bcc metals and fcc Au [9].

The relative difference in strength of the materials decreases with decreasing pillar size. The strength values correlate with the athermal temperature of the metals, with W showing the highest strength values and Nb the smallest ones. The power law exponents obtained for W, Mo, Ta and Nb are -0.21, -0.38, -0.41 and -0.48 respectively. A plot of the power law exponents of the four bcc metals with respect to the ratio of T_{test} to T_{ath} (figure (6.3)) reveals that the power law exponents follow a linear relationship according to

$$\beta = -0.01062 \pm 0.0451 + (-0.57341 \pm 0.06943) \cdot \frac{T_{test}}{T_{ath}}. \tag{6.1}$$

Extrapolating this linear relationship to $\frac{T_{test}}{T_{ath}} = 1$ shows that the power law exponent expected for this value is -0.6 which corresponds to the mean value found for fcc metals. A plot which shows the temperature and size dependence of Nb, Ta, Mo and W is given in the Appendix (A.3).

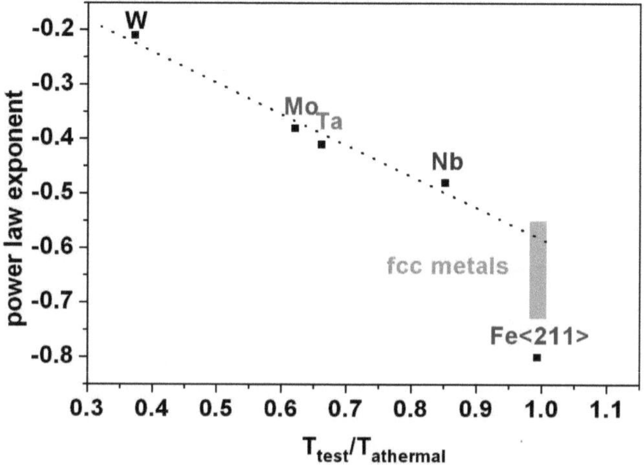

Figure 6.3: Power law exponents of five different bcc metals plotted with respect to $\frac{T_{test}}{T_{ath}}$

The athermal temperature of Fe is approximately at 300K. The exponent observed in α-Fe is added in figure (6.3) to the data from [79] and confirms the trend of stronger size dependence with increasing ratio of T_{test} to T_{ath}. The linear relationship of the power law exponent and the ratio of T_{test} to T_{ath} (equation (6.1)) is not supported by the Fe data. This relationship is based on four metals with a $\langle 001 \rangle$ out-of-plane orientation, whereas the Fe pillars were $\langle 211 \rangle$ oriented. Furthermore Fe is a ferromagnetic material at room temperature. It is known that the magnetic properties of the metal also affect its mechanical properties which may explain the observed differences.

The ratio of T_{test} and T_{ath} is a measure for the thermal activation of the screw dislocations and therefore related to their mobility. Summarizing the comparison of different bcc metals, one can say that a reduced mobility of the screw dislocations leads to higher strength and a less pronounced size dependence.

The flow stresses in figure (6.2) are not normalized. It can be argued that the temperature dependency disappears when the data is normalized in an appropriate way. Two different normalization procedures have been tested. First, the data was normalized with respect to shear modulus μ and, second, the bulk yield stress was subtracted prior to normalizing with μ. Normalizing the data by shear modulus μ will only shift the curves, but does not affect their slope (cf. Appendix (A.4)). The second normalization procedure is shown in figure (6.4) and shows that the temperature dependence of the size effects of bcc metals persists even after this procedure.

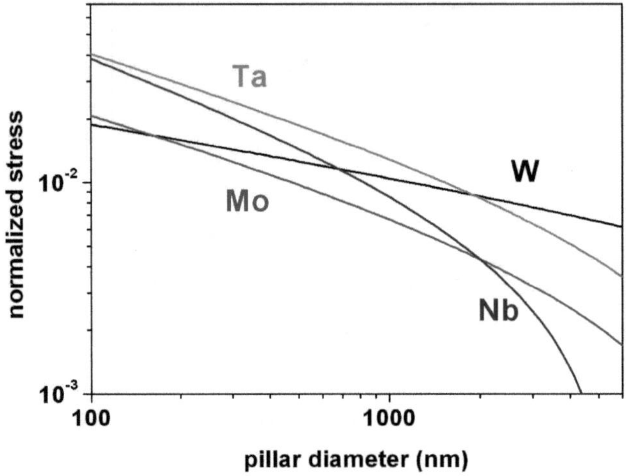

Figure 6.4: Normalized flow stresses of W, Mo, Ta and Nb with respect to pillar diameter. Prior to dividing the flow stress by μ, the bulk yield stress was subtracted (W=550MPa, Mo=430MPa, Ta=345MPa, Nb=240MPa. (values from www.goodfellowusa.com))

In table (6.2) power law exponents of different fcc metals are shown. It is obvious that the exponents of the bcc metals scatter over a wide range whereas the exponents of the fcc metals are concentrated in a rather small area. This is also emphasized when the power law exponents are plotted with respect to the ratio of T_{test} and T_{ath} (figure (6.5)). For fcc metals an athermal temperature is not defined because screw and edge dislocation have almost equal mobilities. Therefore, the ratio of T_{test} and T_{ath} is set to one for the fcc metals.

LATTICE	MATERIAL	POWER LAW EXPONENT N
bcc	Mo⟨001⟩	-0.22 [80]
	Mo⟨235⟩	-0.34 [80]
	Mo⟨001⟩	-0.45 [10]
	W⟨001⟩	-0.21 [79]
	Mo⟨001⟩	-0.38 [79]
	Nb⟨001⟩	-0.48 [79]
	Nb⟨001⟩	-1.06 [11]
	Ta⟨001⟩	-0.42 [this work]
	Ta⟨111⟩	-0.36 [thiswork]
	Ta⟨235⟩	-0.33 [this work]
	Fe⟨211⟩	-0.81 [this work]
fcc	Au⟨001⟩	-0.97 [10]
	Ni⟨111⟩	-0.69 [50]
	Ni⟨269⟩	-0.64 [49]
	Au	-0.61 [9]
	Al	-0.73 [8]
	Cu	-0.59 [8]
	Cu⟨235⟩	-0.55 [this work]

Table 6.2: Power law exponents for different materials and lattices

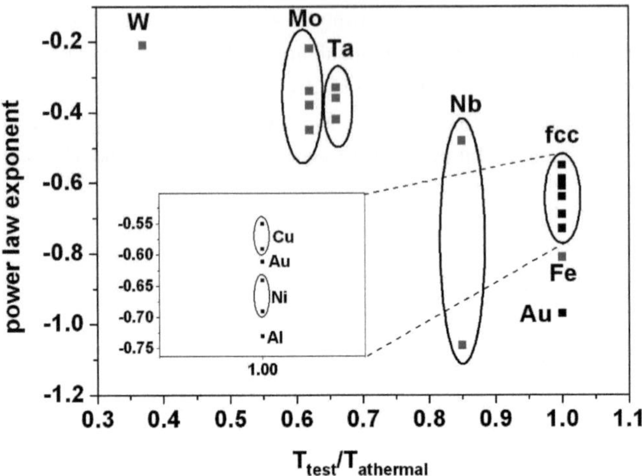

Figure 6.5: Power law exponents with respect to the ratio of room temperature and athermal temperature. BCC metals (red) show a temperature dependency of their exponent, whereas the fcc metal have exponents close to -0.6 (with the exception of Au from reference [10] and Niobium from reference [11])

6.5 Possible Mechanisms Affecting Size Effects in bcc Metals

The comparison of fcc and bcc metals and the comparison of different bcc metals among each other indicates that mechanical size effects do not have the same underlying mechanisms in fcc and bcc metals. Besides the mechanisms mentioned in chapter (2.4.3), additional mechanisms may be active in bcc metals.

6.5.1 More Dislocation Sources in BCC

The complex nature of screw dislocations in bcc metals leads to a higher dislocation source density as compared to fcc metals. The dislocation multiplication processes in bcc metals are discussed in detail in chapter (2.3.2). These processes may account for the difference in size dependent behaviour which is observed between fcc and bcc metals.

Immobile dislocation segments can act as pinning points for dislocation multiplication. The probability to form immobile segments is higher in bcc than in fcc metals which will be illustrated in the following. Dislocation sources may form by the double cross slip-mechanism (cf. chapter (2.3.2)), where relatively immobile edge segments form and act as pinning points. This process is more likely to occur in bcc than in fcc systems because of two reasons. First, in bcc metals more glide planes are available than in fcc metals: Screw dislocations with $\vec{b} = \frac{1}{2}\langle 111\rangle$ can glide on three $\{110\}$ and three $\{211\}$ planes whereas in fcc metals a screw dislocations with $\vec{b} = \frac{1}{2}\langle 110\rangle$ can only glide on two planes of the $\{111\}$ type. Second, in bcc metals the motion of dislocations naturally involves more cross slip. The motion of a dislocation on a glide plane is not only determined by the resolved shear stress but also by the mobility of the dislocation [24]. One example for this is the frequently observed "wavy-slip" which is periodic cross slip between two planes. Another mechanism that can lead to immobile dislocations is based on kink reactions. Screw dislocations move by kink pair-nucleation. Kinks can form on different glide planes and when moving along a screw dislocation, they can interact and become immobile. This will result in a pinning point where further kinks can pile up and form dislocation debris loops. This observation was made using discrete dislocation dynamics simulations [4]. The mechanisms discussed above lead to an increased dislocation source density in bcc metals as compared to fcc metals and may have effects on the size dependence of the mechanical behaviour. For very small samples where dislocation starvation is important, dislocation nucleation may be alleviated due to illustrated higher source density in bcc metals. This may shift the onset for starvation to smaller sample sizes and may lead to a weaker size dependence of the yield strength for bcc metals.

6.5.2 Surface Related Effects

The volume of a body scales with x^3 with a characteristic dimension x whereas its surface scales with x^2. If the sample size is reduced, the ratio of surface to volume gets larger. This means the deformation behaviour of small scaled samples is more affected by the surface. Image stresses exert forces on dislocations which affect their motion (cf. chapter (2.3.3)).

In the vicinity of the surface, image stresses or stress concentrations due to surface roughness exist. These stresses can bend dislocations and facilitate nucleation of kinks at the surface. In bcc metals, the mobility of screw dislocations is limited by the kink pair-nucleation mechanism. Edge components close to the surface can supply kinks to the screw dislocations and thereby enhance their mobility. The increased mobility results in a reduced strength of the material. Since the surface to volume ratio increases with decreasing sample size, this mechanisms leads to a relative mechanical weakening when the sample size decreases which may contribute to the weaker size dependence of bcc metals that is shown in figure (6.5). Also, the activation volume (equation (6.10)) of small samples would be expected to be reduced compared to bulk specimens. The influence of the surface on the screw dislocations is discussed in more detail in chapter (6.6).

6.5.3 Kinetic Pile-up

Since screw dislocations have a lower mobility than edge dislocations, pile-ups of screw dislocations may occur [80]. In bcc pillars, this could lead to a configuration of dislocations as shown in figure (6.6) which is similar to the one discussed by Vitek and Groeger [100]. They explained the apparent low Peierls stresses that were measured in the literature by pile-ups of screw dislocations that help moving individual dislocations through the crystal. When a source emits dislocations, the dislocations first have a high curvature but when they move away from the source their curvature will decrease. During this process, edge dislocations will move faster than the screw segments. As a result, elongated screw segments form that exert a back stress and eventually limit the motion of the dislocation loops. The straight screw segments pile up in dislocation arrays until the back stress on the source attains stresses where the source cannot be activated anymore. In small columns, where the deformation may be source limited, the size of the array of screw dislocations and its back stress on the source depend on the column dimensions. Smaller columns contain a smaller number of dislocations and the source can be activated more easily. This mechanism also could alleviate the effect of dislocation starvation and may account for the weaker size dependence slope that is observed in bcc metals. Also, the pile-up of dislocations to arrays may account for the higher values for the strengths that are found in Ta samples of several micrometers in size. In this regime dislocation interaction may dominate and the interaction of arrays may lead to higher stresses than the interaction of individual dislocations.

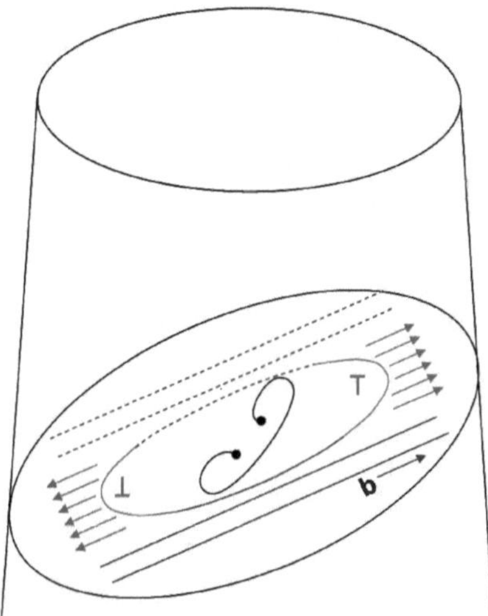

Figure 6.6: Pile up of screw dislocations in the vicinity of an operating dislocation source.

6.6 Microcompression Applied to Cuboidally Shaped Pillars

As can be seen from the experimental data (chapters (5.1.2), (5.2.3) and (5.3.3)), the microcompression of pillars in the SS configuration resulted in higher strain values than in the LS configuration in the majority of cases. This difference in strain has its origin in different true stresses of the two orientations during compression. Figure (6.7) shows the two configurations after a displacement of Δx. This displacement leads to a change of area ΔA which can be determined as follows.

$$\begin{aligned} \Delta A_{LS} &= \Delta x \cdot a \\ \Delta A_{SS} &= \Delta x \cdot c \\ \rightarrow \frac{\Delta A_{SS}}{\Delta A_{LS}} &= \frac{c}{a} \end{aligned} \tag{6.2}$$

The change of area is larger by a factor of $\frac{c}{a}$ for the SS than for the LS configuration which leads to higher true stresses during deformation in the SS configuration. After the onset of plasticity these higher true stresses lead to an abrupt deformation of the pillar, so that the indenter overshoots the 10% strain at which compression is supposed to stop. It can be argued that the higher stresses observed for the LS configuration in Ta are a consequence of the higher true stresses in the SS configuration,

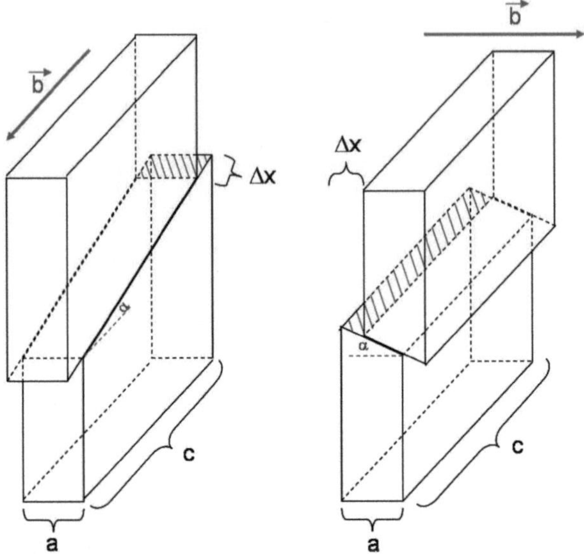

Figure 6.7: The two different orientations lead to different changes of cross section during deformation. The change of area of the LS configuration (left) exceeds the change of area of the SS configuration (right) by a factor $\frac{c}{a}$.

since only engineering stresses were considered (cf. chapter (5.2.3)). This issue has been investigated and it was found that the effect still exists when the true stresses are taken into account. The difference in flow stress was found to be approximately 40MPa. The 'true-stress'-model neglects elasticity of the material and assumes slip on only one glide plane. It therefore represents a worst case scenario that has never been observed in the experiments. The higher stresses observed for the LS configuration in Ta are therefore not only a geometrical effect, but also have their origin in dislocation processes. Details of this treatment of this issue are given in the Appendix (A.5).

The experiments on cuboidal pillars were performed in order to investigate the role of screw dislocations in bcc metals. As already mentioned there is a discrepancy of the mobilities of screw and edge dislocations when bcc metals are exposed to shear stresses. Mixed dislocations will therefore turn into pure screws since the edge fractions will leave the sample comparatively fast (cf. chapter (6.5.3)). This will lead to an arrangement of elongated screw dislocations. The alignment of the pillars with respect to the Burgers vector determines the length of these screw dislocations. The sample alignment procedure was discussed in detail in chapter (3.5). The alignment

parallel to the Burgers vector can be considered as long screw orientation (LS) whereas the perpendicular configuration will result in short screw segments (SS). The expected dislocation structure is illustrated in figure (6.8). The two orientations are supposed to affect the deformation behaviour of the pillar as long as bcc metals below their athermal temperature are considered. If screw and edge dislocations have comparable mobilities like it is the case in fcc metals or bcc metals above their athermal temperature, the deformation behaviour should not depend on the in-plane orientation of the cuboidally shaped pillars. Ta is the only material tested that shows higher flow stresses for the LS configuration than for the SS configuration. Experiments on cuboidally shaped Mo pillars [101] at room temperature show higher flow stresses in the LS configuration as well which is in good agreement with the observations made in this work. Fe and Cu show the same flow stresses for both orientations. This behaviour is not surprising, since edge and screw dislocations have approximately equal mobilities in these metals. The effect of higher flow stresses in the LS configuration is only observed when screw dislocations require thermal activation in order to move. These observations support the theory that size dependent mechanical behaviour of bcc metals is affected by the temperature dependent screw dislocation mobility.
The orientation dependence of Ta can be explained by interactions of the screw dislocation with the sample surface. Defect dynamics simulations show that the motion of screw dislocations in bcc metals is controlled by single kink-nucleation from the surface [5]. Dislocations encounter stress fields in the vicinity of the surface. These stress fields arise because of surface roughness or image stresses and affect the dislocations in the following manner. Screw dislocations experiencing the stress fields at the surface can be bent towards the surface. This will turn the pure screw dislocation into a mixed dislocation with the edge fragments having a high mobility. These edge segments serve as kinks and the thermal activation required to move the dislocation is reduced. This effect is more effective in the SS configuration than in the LS configuration because in the SS case a larger fraction of the screw dislocation experiences the surface stresses.

Furthermore, the stress fields at the surface enhance the self multiplication of the screw dislocations (cf. chapter (2.3.2)). This multiplication process is more likely to occur in the SS configuration because more screw dislocations will be present in the sample and for a given radius of bending the short segments are stronger affected by the surface stresses than the long segments.
Vesely [83, 102] observed in a study on Mo single crystals that image stresses in the vicinity of the surface affect the selection of the slip systems that will be activated when the sample is deformed. In this study anomalous slip was observed and therefore a criterion was developed which allows the determination of the activated slip systems.

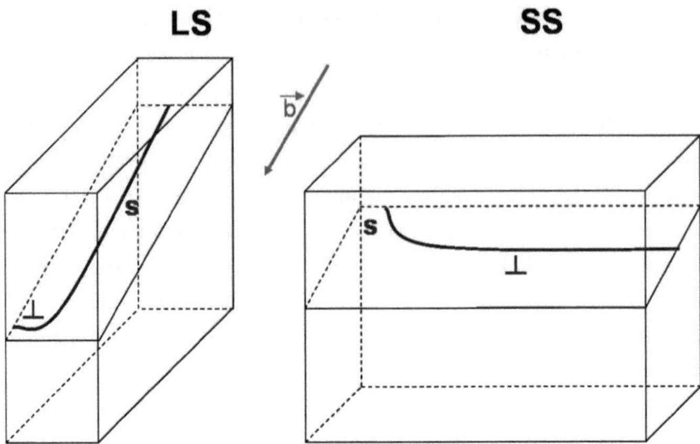

Figure 6.8: Mixed dislocation inside a cuboidal pillar. The length of the screw and edge components depends on the orientation of the pillar with respect to $\vec{b}$.

This criterion corresponds to Schmid's law with an additional factor sinβ, with β being the angle between the Burgers vector (respectively the line vector of a screw dislocation) and the normal vector of the sample surface.

$$\rightarrow V = \sin\beta \cdot \underbrace{\cos\phi \cdot \cos\lambda}_{\text{Schmid factor}} \tag{6.3}$$

The glide system with a maximal value of V will be activated.
Matsui et al. developed a model for this surface effect [12] and observed this effect experimentally [103]. This mechanism is illustrated in figure (6.9).
Figure (6.9(a)) The screw dislocation OO' is lying in the plane of the paper, whereas the surfaces S and S' of the specimen are perpendicular to it. Image forces act on the dislocation line and will deflect the segment OA to FA. The same effect will happen on the reverse side S' of the specimen when no stresses are applied on the specimen.
Figure (6.9(b)) A shear stress is applied on the sample which causes the dislocation to move downwards. The mixed dislocation segment FA will move downwards to F_1A_1. On the reverse side of the specimen the shear stress opposes the image stress and deflects the segment A'F' back to its initial configuration F'O'. If the shear stress is large enough, the downward movement of the segment F_1A_1 will continue.
Figure (6.9(c)) The difference in mobility of screw and edge components will change the shape of the mixed segment gradually from 1 to 5. This segment traverses along the original screw dislocation as a large kink as illustrated in figure (6.9(d)). This surface effect considerably reduces the stress, that is required to move the dislocation.

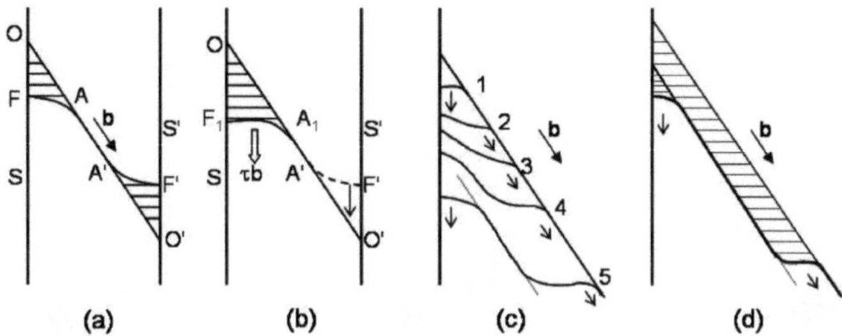

Figure 6.9: Image stresses bend the screw dislocation line and therefore enhance their mobility. Figure taken from [12].

The surface effect may also trigger dislocation sources which is illustrated in figure (6.10). The primary slip plane KLMN and the anomalous slip plane GHIJ are both of the $\{110\}$ type. The planes intersect at the straight line OO', which is a $\langle 111 \rangle$ direction. At a stress slightly smaller than the yield stress, a ledge at the surface of the sample may act as dislocation source and introduces the half loop OEP in the primary slip plane. The edge segment E moves through the crystal until it faces an obstacle, for instance inclusions or other immobile dislocation segments. The screw segments OE and PE are not able to move by the double kink mechanism at this stress. The image stresses at the surface attract the segment OA and the segment is deflected to a mixed segment FA in the anomalous plane. This mixed segment moves along AE in the anomalous plane and therefore leads to motion of the whole dislocation line OE in the anomalous plane. When the end of the obstacle is reached, the edge segment will move through the crystal and form a new dislocation F_2F_2' (figure (6.10(b))). The remaining dislocation segment EF' will now move upwards since the sense of the dislocation has been reversed. The segment EF' continues to turn around about the pole dislocation E in the anomalous plane as shown in figure (6.10(c)). The surface effect therefore leads to a self multiplication of the screw dislocations and to anomalous slip which is often observed in bcc metals.

The multitude of possible dislocation multiplication mechanisms (cf. chapter (2.3.2)) leads to the conclusion that dislocation starvation does not play a role when mechanical size effects in bcc metals are considered. Strengthening by dislocation starvation in small bcc samples becomes even more unlikely when dislocation sources can be generated by stress concentrations in the vicinity of a surface since the surface to volume ratio increases with decreasing sample dimensions.

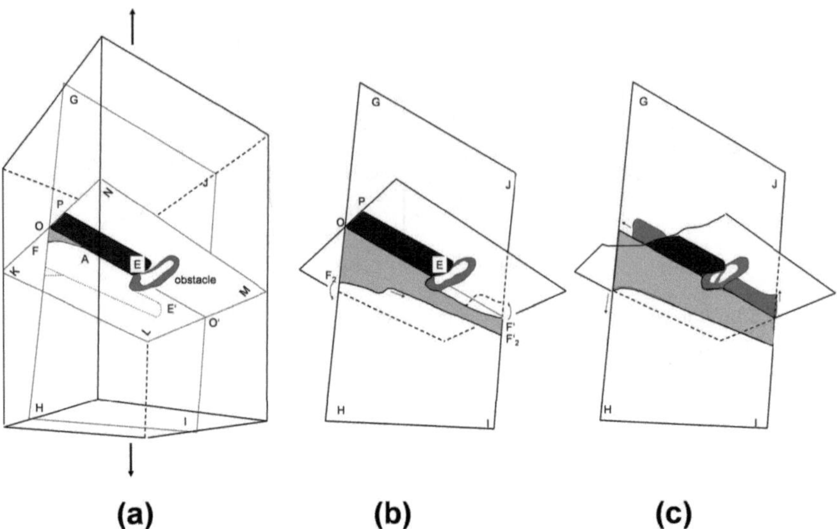

Figure 6.10: Dislocation multiplication triggered by the surface effect [12].

6.7 Comparison of Cylindrical and Cuboidal Pillars

In a study on cuboidal W pillars [81] it is shown that the smallest dimension of the pillar determines its yield strength. The experiments made on cuboidal pillars in this work were also compared to experiments made on cylindrical pillars. The yield strength of the cuboidal Ta pillars depends on the in-plane orientation of the pillars (cf. chapters (5.1.2), (5.2.3) and (5.3.3)). To investigate whether the small dimension of the cuboidal pillars dominates the deformation behaviour, the 5% flow stresses of the circular and cuboidal pillars were plotted into the same diagram. On the abscissa the diameter of the cylindrical respectively the small edge length of the cuboidal pillars is plotted (figure (6.11) and (6.12)). This treatment of the data shows that the 5% flow stress of the Cu pillars is determined by the smallest dimension of the pillars. In Ta the 5% flow stress of the cuboidal pillars is not only a function of the smallest dimension, it also depends on the alignment of the pillars. The LS configuration shows higher stresses than the SS configuration as is discussed in chapter (6.6). The cylindrical Ta pillars show the same flow stresses as the cuboidal pillars with the SS orientation. This means that the flow stresses of the cylindrical pillars correspond to the lower limit of the cuboidal pillars and do not show a mixed behaviour of SS and LS configuration. When considering dislocation multiplication and motion it is also possible that the yield strength of a pillar depends on surface area, volume or surface to volume ratio.

The values of these parameters are compared in table (6.3). For the cuboidal pillars the geometry is assumed to be $3 \cdot 1 \cdot 5 \mu m^3$ (length· width· height), the cylindrical pillars have a diameter at midheight of $1\mu m$, a height of $3\mu m$ and a taper angle of $3°$.

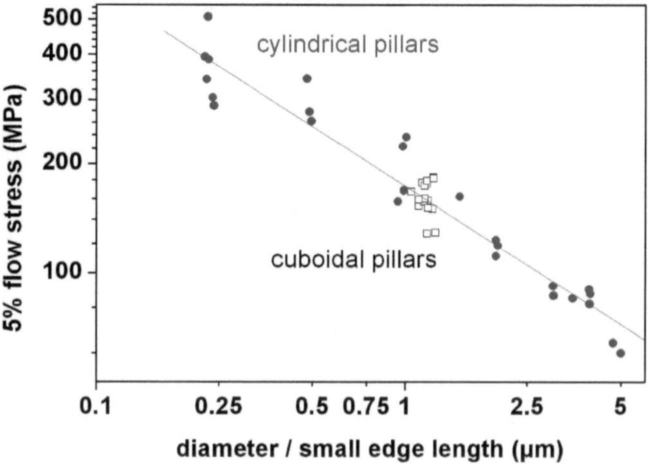

Figure 6.11: Comparison of cylindrical and cuboidal Cu pillars. The flow stress at 5% plastic strain is plotted versus the diameter of the cylindrical pillars and the smallest dimension of the cuboidal pillars, respectively.

PARAMETER	CUBOIDAL PILLARS	CYLINDRICAL PILLARS
surface area A	$43\mu m^2$	$9.97\mu m^2$
volume V	$15\mu m^3$	$2.4\mu m^3$
$\frac{A}{V}$	$2.87\frac{1}{\mu m}$	$4.15\frac{1}{\mu m}$

Table 6.3: Comparison of surface area, volume and surface to volume ratio of cylindrical and cuboidal pillars

Plotting the 5% flow stress with respect to surface area, volume or surface to volume ratio would not bring the data points of the cuboidal and cylindrical pillars together, which indicates that flow stress is rather a function of the smallest dimension than of surface area, volume or surface to volume ratio. This behaviour can be explained by a critical stress that is required to move a dislocation. The activation of dislocation

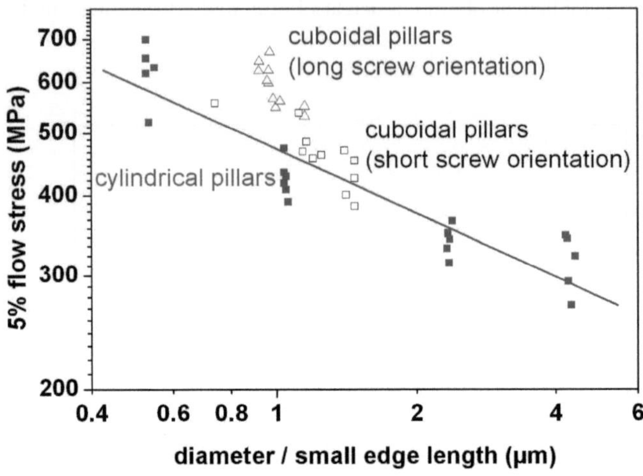

Figure 6.12: Comparison of cuboidal and cylindrical Ta pillars. The flow stress at 5% plastic strain is plotted versus the diameter of the cylindrical pillars and the smallest dimension of the cuboidal pillars, respectively. Long screw and short screw orientation are according to chapter (3.5).

sources seems to play a less important role since the number of dislocation sources is supposed to correlate with the volume of the sample. A larger volume has a higher probability of containing sources that can be easily activated, therefore smaller stresses can be expected for a larger volume if the deformation was controlled by the activation of dislocation sources. The comparison of the cylindrical and cuboidal pillars shows that dislocation starvation does not seem to be active at the size scale that is considered here. For the first time, it has been observed that increasing sample size (in a certain direction) leads to strengthening (Ta in the LS orientation).

6.8 An Attempt to Describe the Surface Effect in a Composite Material Model

The experiments that were performed on cuboidally shaped pillars strongly indicate that the surface enhances the mobility of screw dislocations in bcc metals. The comparison of the mechanical size effects of fcc and bcc metals shows that bcc metals exhibit higher stresses and show a weaker size effect, even when normalized stresses are considered (chapter (6.3)). The difference in stress between fcc and bcc metals can be attributed to the low mobility of screw dislocations. As the samples become smaller the influence of the surface becomes stronger, leading to an enhanced mobility of the screws in the proximity of the surface. This surface effect counteracts the mechanical size effect and may be the origin of the reduced size dependence of bcc metals. The relative difference between the stresses of the fcc and bcc metals becomes smaller as the sample dimensions are reduced and the stresses become equal at very small sample sizes ($\approx$ 100nm). One may assume that the stresses are equal when the whole cross section of the bcc pillar is affected by the surface and the screw dislocations do not require thermal activation. The zone of influence of the surface is on the order of 100nm as can be estimated from TEM-investigations on Mo samples [103]. This will lead to equal stresses of fcc and bcc pillars for pillars with a diameter around 200nm. The mechanical behaviour of larger pillars may then be considered to be a superposition of an fcc component and an additional component that is attributed to the low mobility of the screw dislocations. Therefore, the fcc term dominates plasticity in the proximity of the surface whereas the inner part of the pillar is dominated by the low mobility of the screw dislocations. The flow stresses σ_{bcc} of the bcc pillars can be described by a superposition of fcc and 'low screw dislocation mobility' behaviour according to

$$\frac{\sigma_{bcc}(d)}{\mu_{bcc}\cdot b_{bcc}} = f\cdot\underbrace{\frac{\left(\sigma_{0fcc}+\sigma_0\cdot\left(\frac{d}{d_0}\right)^{-\beta}\right)}{\mu_{fcc}\cdot b_{fcc}}}_{\text{fcc behaviour}} + (1-f)\cdot\underbrace{\frac{\left(\sigma_{0bcc}+\tilde{\sigma_0}\cdot\left(\frac{d}{d_0}\right)^{-\beta}\right)}{\mu_{bcc}\cdot b_{bcc}}}_{\text{'low screw dislocation mobility' behaviour}} \tag{6.4}$$

with f the cross-sectional fraction of the pillar that is affected by the surface effect, d the pillar diameter, d_0 characteristic length-scale (in the following $d_0 = 1\mu\text{m}$), b the Burgers vector, μ_{fcc} and μ_{bcc} the shear moduli of the fcc and bcc metal, σ_{0fcc} and σ_{0bcc} the bulk yield stress of the fcc respectively the bcc metal, σ_0 and $\tilde{\sigma_0}$ constants and β the power law exponent. The power law exponents of the surface area and the interior part of the pillar are supposed to be equal, since it is assumed that both parts of the pillar obey the same dislocation source statistics according to the Orowan equation. The Orowan equation gives the flow stress $\sigma_{f,source}$ of a thin metal film with respect to the size of the dislocation source q:

$$\sigma_{f,source} = \frac{1}{s} \cdot \frac{\mu b}{2\pi} \cdot \frac{1}{q} \ln\left(\frac{\alpha' q}{b}\right) \tag{6.5}$$

where s is the Schmid factor of the corresponding slip system, α' is a constant, b is the Burgers vector, and μ is the shear modulus of the film material [38]. The source size increases with increasing film thickness and the flow stress is reduced. For pillars, the flow stress decreases as the diameter of the pillar increases.

For cylindrical pillars the fcc component is a cylindrical shell and f is determined to be

$$f = \frac{4 \cdot (a \cdot d - a^2)}{d^2}. \tag{6.6}$$

Like mentioned above, the zone that is affected by the surface is on the order of 100nm, and therefore value of 100nm was used for a.

This simple model was applied to Ta whereas Cu was used as a reference fcc metal. The normalized flow stresses of the fcc metals show a common scaling behaviour, so that Cu can be considered to exhibit typical fcc behaviour (figure (2.11)). The scaling of Ta and Cu was obtained from figure (5.9) and (5.3). For the bulk yield stresses σ_{bulk}, Burgers vectors b and shear moduli μ of Cu and Ta the following values were found in the literature, Cu: σ_{bulk}=26MPa [104], μ_{Cu}=48GPa, b=0.2556nm and Ta: σ_{bulk}=124MPa [13], μ_{Ta}=69GPa, b=0.285nm.

$$\text{Cu: } \sigma = 26\text{MPa} + 143 \cdot \left(\frac{d}{d_0}\right)^{-0.67} \text{MPa} \tag{6.7}$$

$$\text{Ta: } \sigma = 124\text{MPa} + 342 \cdot \left(\frac{d}{d_0}\right)^{-0.47} \text{MPa} = \sigma_{bcc}(d) \tag{6.8}$$

Equations (6.7) and (6.8) can be inserted into equation (6.4), so that there is only one unknown variable left, namely $\tilde{\sigma_0}$. This variable can be determined using an arbitrary value (in this work: $\sigma_{bcc}(2\mu\text{m}) = 371\text{MPa}$).

$$\frac{\sigma_{Ta}(d)}{69\text{GPa} \cdot 0.285\text{nm}} = f \cdot \frac{\left(26\text{MPa} + 143 \cdot \left(\frac{d}{d_0}\right)^{-0.67} \text{MPa}\right)}{48\text{GPa} \cdot 0.2556\text{nm}} + (1-f) \cdot \frac{\left(124\text{MPa} + 462 \cdot \left(\frac{d}{d_0}\right)^{-0.67} \text{MPa}\right)}{69\text{GPa} \cdot 0.285\text{nm}} \tag{6.9}$$

At this point it is not proven that the model works, but it already meets certain conditions, $\lim\limits_{d\to\infty} = \sigma_{bulkTa}$ and for d$\leq$ 200nm $\rightarrow \frac{\sigma_{Ta}(d)}{69\text{GPa}\cdot 0.285\text{nm}} = \frac{(26\text{MPa}+143\cdot d^{-0.67}\text{MPa})}{48\text{GPa}\cdot 0.2556\text{nm}}$ (pure fcc behaviour).

Equation (6.9) allows the determination of the flow stresses of the cuboidal pillars. The shortest dimension of the cuboidal pillars determine the flow stress as has been shown by the study of Jennett et al. [81] and the comparison of cylindrical and cuboidal pillars in this work in chapter (6.7). The mean dimensions, flow stresses and the resultant values for f of the cuboidal pillars are shown in figure (6.13).

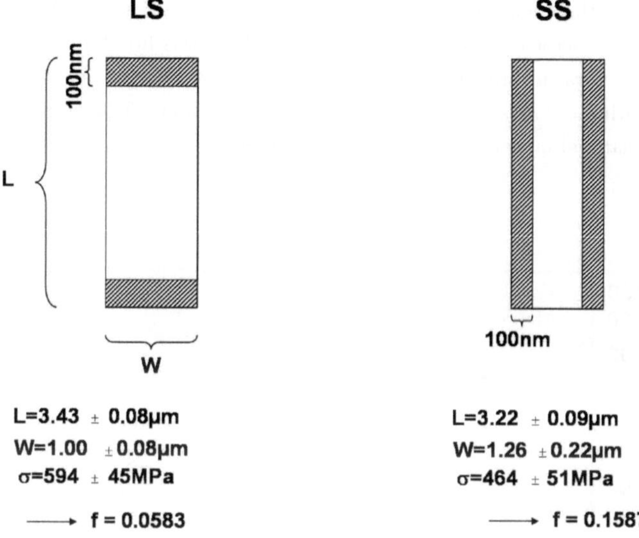

Figure 6.13: Surface effect in cuboidal pillars. The mean dimensions (length L and width W) and the measured flow stresses are given.

Using the values of the cuboidal pillars in equation (6.9) gives stress values close to the expected ones for the SS and LS configuration.

CONFIGURATION	LS	SS
flow stress from measurement	594 ± 45MPa	464 ± 51MPa
flow stress from calculation	568MPa	475MPa

Table 6.4: Measured and calcualted stresses for the cuboidal pillars in LS and SS configuration

Another condition that needs to be fulfilled is that the flow stress of the cuboidal pillars has to be independent of L. This condition is a consequence of the fact that the shortest dimension of the cuboidal pillar determines its strength as has been shown

by [81] (chapter (2.4.3)) and the experiments performed on cuboidally shaped Ta- and Cu-pillars in this work (chapter (6.7)). If L is increased, the value of f is decreased in the LS configuration. The flow stress becomes higher and reaches its maximum of 586MPa for $f = 0$, the increase in stress is almost negligible. For the SS configuration the value of f remains constant with increasing L, so that the flow stress is solely determined by the shortest dimension of the cuboidal pillar.
The results of the composite material model are shown in figure (6.14). The data obtained from microcompression of cylindrical pillars, was fitted with the functions (6.7) and (6.8)(drawn through lines). The calculated stresses of the cuboidally shaped pillars lie within the area of confidence for both orientations. This indicates that the composite material model works well in this particular case.

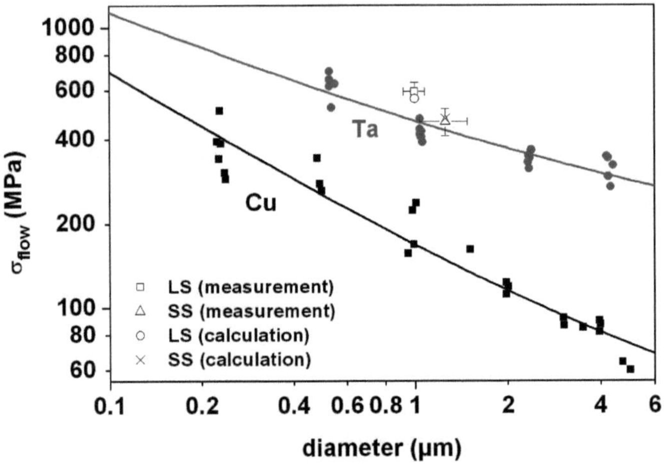

Figure 6.14: Size effect of Cu and Ta with the measured and calculated values of the cuboidally shaped pillars.

6.9 Strain Rate Sensitivity of Small Ta and Fe Pillars

The results of the strain rate sensitivity and activation volume analysis are shown in figure (6.16). Here, the flow stresses at 5% plastic strain are plotted as a function of the plastic strain rate, as determined from an interval of 10 data points (respectively 1s) at 5% plastic strain (figure (6.15)). This procedure is necessary since the strain rate is not constant throughout the measurement. The error bars in flow stress (figure (6.16)) contain the limited resolution of the nanoindenter and uncertainties in the measurement of the column diameter. The latter was estimated to be 10% of the middle column diameter. The error bars in strain rate represent the maximum deviation of the strain rate within the interval in which the strain rate was calculated (figure (6.15)). From the strain rate analysis an activation volume has been calculated according to

$$\nu^* = mk_BT\left(\frac{\partial ln\dot{\epsilon}}{\partial\sigma}\right) \tag{6.10}$$

with m = Taylor factor, k_B = Boltzmann constant, T = temperature, $\dot{\epsilon}$ = strain rate and σ = stress [26, 80, 105–107].

In figure (6.16), the solid line represents the linear regression according to equation (6.10) with the slope of the regression line being inversely proportional to the activation volume. The dashed lines indicate the area of confidence of the line of best fit. For comparison, values for bulk material [13] are included in the diagram. An activation volume of $5.7b^3$ for the 4µm-columns with $10.8b^3$ and $3.9b^3$ as upper and lower confidence limits was obtained. As can be seen from figure (6.16), the upper limit corresponds to the activation volume measured for bulk Ta. The activation volume is not a physical volume, but it is related to the area swept by a dislocation in an activation event [108]. The activation volume for deformation of bcc metals is in the range of $5b^3$ to $100b^3$. The activation volume of fcc ranges from 10 to 100 times larger [13]. The activation volume found for Ta is in the range that can be expected for bcc metals. The small activation volume suggests that the deformation energy has been dissipated in a small activation site involving only a few atoms, i.e. the nucleation of kinks. The somewhat smaller volume determined here, compared to the bulk value, may be related to the formation of kinks at the surface compared to kink pairs in the bulk. In fcc metals the energy is dissipated along longer sections of the dislocation line which leads to a much larger activation volume.

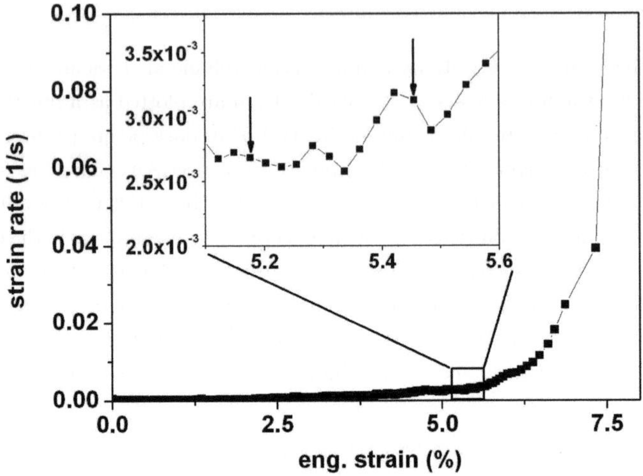

Figure 6.15: Representative data for strain rate vs. strain, from which the strain rate at 5% plastic strain was determined. As the test is performed in load control the strain rate increases when plastic flow sets in. In the inset the data close to 5% plastic strain (5.3% total strain) is enlarged for clarity. A strain burst at approx. 7.5% can be observed. The arrows mark the interval from which the plastic strain rate was calculated.

The load rate dependence of Fe was investigated by means of microcompression and nanoindentation. As can be seen in figure (5.26) and in figure (5.27) the results of the microcompression experiments are not unambiguous. The 0.5µm pillars do not exhibit a load rate dependent behaviour whereas the 1µm pillars show an increase in the 5% flow stress with higher loading rate. The stochastical occurrence of strain bursts during the compression of pillars at these sizes leads to a non-negligible error that may cause the different results. For clarification, nanoindentation experiments were carried out. The results are shown in figure (5.29). 45 indents were performed, 15 for each indentation strain rate. The averaged force-displacement curves show a significant dependence on the strain rate. This is in agreement with the work of Ostwaldt et al. [109] who report that there is a strain rate-sensitivity of Fe at small strain rates. Ostwaldt et al. did not investigate the underlying mechanisms of the strain rate sensitivity at rates smaller than $10^2 s^{-1}$. At rates higher than $10^2 s^{-1}$ strain rate sensitivity can be attributed to the formation of twins. In a study by da Silva et al. [110] compression

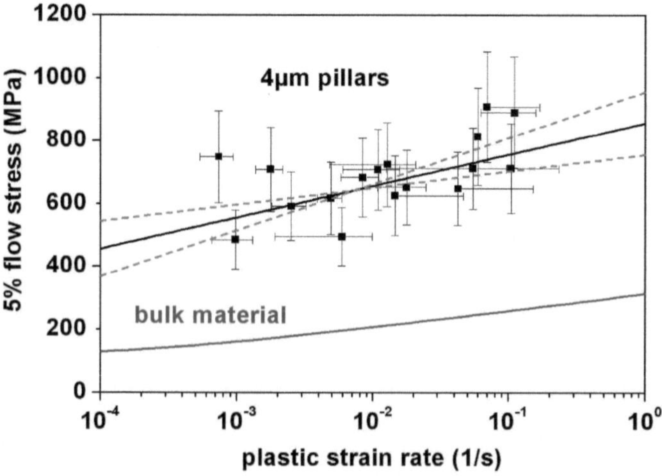

Figure 6.16: Strain sensitivity of Ta: Flow stress at 5% plastic strain as a function of the plastic strain rate for pillars with a diameter of 4μm. The regression line (solid) is a measure for the inverse activation volume according to equation (6.10). The dashed lines confine the area of confidence. For comparison, the behaviour of bulk Ta is included according to [13].

experiments on Fe were performed at strain rates between $10^{-4}s^{-1}$ and $10^{4}s^{-1}$. A strain rate dependence of the flow stress of Fe was observed as well, so that the absence of a strain rate sensitivity in the 0.5μm pillars may be attributed to their stochastical deformation behaviour. The experiments were carried out at room temperature, where screw dislocation motion does not require thermal activation in Fe. So, the presence of a strain rate sensitive deformation behaviour may be explained by other mechanisms that require thermal activation like climbing of dislocations to circumvent impurities or immobile dislocation segments.

7

Summary

Microcompression testing has become a commonly used technique to investigate mechanical properties of small-scaled samples. The main goal of this work was to investigate small-scale plasticity of bcc metals. Beyond the investigation of bcc metals, microcompression could be successfully used to obtain Young's moduli of Silicon for different crystallographic directions.

From Si⟨111⟩ and Si⟨100⟩ micropillars with different aspect ratios were machined. The microcompression experiments were performed in the elastic regime of the Si-pillars. It was observed that pillars with high aspect ratio exhibit a higher Young's modulus which can be attributed to substrate compliance. The ⟨111⟩-direction exhibited higher values for the Young's modulus than the ⟨100⟩-direction. In addition to the experiments FEM-simulations of the microcompression tests were carried out. Experiments and simulations were in good accordance with each other, which shows that it is possible to measure the Young's modulus of individual crystallographic orientations by means of microcompression.

Microcompression in combination with EBSD offers the possibility of identifying activated slip systems. A technique was developed which allows the identification of slip systems in cylindrical and cuboidal pillars. This glide plane identification was performed on cylindrical and cuboidal Ta pillars. Furthermore the identification of slip systems was revised on cuboidal Cu pillars. The results of the slip system identification were compared to the Schmid factors and the results show that the identified slip systems were among the systems with the highest resolved shear stresses. It has been shown, that in bcc metals glide does not necessarily occur on the system with the highest Schmid factor. This can be rationalized as the Schmid factor is a value that only takes geometrical parameters into account. Packing densities of the different glide planes, local stress concentrations and effects from the core splitting of the screw dislocations are not considered by the Schmid factor.

Microcompression of bcc metals was performed in order to investigate their small scale behaviour and to identify the underlying mechanisms of an occuring mechanical size effect. Single crystalline samples of the bcc metals Ta and α-Fe were tested and both metals exhibited size dependent behaviour. Both metals increase significantly in strength when at least one of the sample dimensions is reduced into the micrometer regime or below. This increase in strength can be quantified by using a power law. The power law exponent is a measure for the increase in flow stress with decreasing sample size. The mechanical size effect in α-Fe is significantly stronger than the one observed for Ta. The comparison with other microcompression studies of bcc metals shows that bcc metals do not exhibit a common scaling behaviour as it is observed for fcc metals. The power law exponent of fcc metals is approximately -0.6. Different normalization procedures have been applied to the data of Ta in order to trace the bcc deformation behaviour back to dislocation mechanisms that are active in fcc metals. It was not possible to normalize the data in a way that fcc metals and Ta fit together. This indicates that there are fundamental differences in the underlying mechanisms of the mechanical size effects in fcc and bcc metals. Comparing the Ta data to three other bcc metals (W, Mo and Nb) reveals that the size effect gets weaker as the ratio of test temperature and athermal temperature of the material gets smaller. At room temperature Nb shows a mechanical size effect that is comparable to the one of the fcc metals whereas the size dependent behaviour of W is much weaker. Although the tested Fe sample had a different orientation as the other four bcc metals, it confirms the observation of the temperature dependent power law exponent.

Screw dislocation motion is thermally activated in bcc metals. The fact that size dependent behaviour and screw dislocation motion both depend on temperature leads to the conclusion that screw dislocations may play an important role in the deformation of small-scaled bcc samples. The role of screw dislocations in small scale plasticity was investigated by microcompression applied to cuboidally shaped pillars. The cuboidal pillars were aligned in a way that either long or short screw dislocation segments are required for deformation. These experiments were performed on Ta, α-Fe and Cu as an fcc reference material. The experiments showed that for Ta higher stresses are required to deform the pillars oriented for long screw dislocations than to deform the pillars oriented for short screw dislocations. No effect was observed for α-Fe and Cu. This is not surprising since for these two metals screw dislocation motion does not require thermal activation at room temperature. In Ta screw dislocations move by thermally activated double kink nucleation. The observed difference in stress indicates that short screw segments are more mobile than long screw segments. Image stresses in the proximity of the surface may bend the screw dislocation line which turns into a mixed dislocation. The edge fragments propagate along the screw dislocation line and enhance its mobility. The sample surface acts as an effective source of kinks. In the short screw configuration a larger fraction of each individual screw dislocation line is affected by the surface than in the long screw configuration. Furthermore it can be found in literature that image stresses may trigger new dislcation sources. This surface effect leads to a higher gain in mobility of the screw dislocations and to additional dislocation sources in the short screw orientation which may explain the difference in flow stress. These experiments strongly indicate that screw dislocation mobility plays an important role in small-scale plasticity of bcc metals.

A comparison of the flow stresses obtained from experiments on cylindrical and cuboidal pillars leads to new insights in the deformation behaviour of small scaled samples. The strength of a cuboidal pillar is determined by the shortest dimension of its geometry. For instance, a cylindrical pillar with a diameter of 1µm and a height of 3µm exhibits the same 5% flow stress as a cuboidal pillar with the following dimensions, $3 \cdot 1 \cdot 5\mu m^3$ (length·width·height). The cuboidal and the cylindrical pillar differ signifcantly in surface area, volume and surface to volume ratio. The only dimension these samples have in common is the smallest dimension, the diameter respectively the small edge length. This indicates that the activation of source dislocations does not seem to affect the stress that is necessary to deform the sample since the number of sources is supposed to correlate with the volume of the sample. The motion of screw dislocation in combination with surface related effects may lead to an enhanced number of dislocation sources by self multiplication of screw dislocations. The flow stress is determined by a critical stress that is needed to move the dislocations. In Ta where the flow stress of the cuboidal pillars is orientation dependent, the flow stresses of the cylindrical pillars are in good accordance with the softest cuboidal pillars. The multiplicity of dislocation sources puts the frequently discussed dislocation starvation mechanism for bcc metals into doubt.

Appendix A

A.1 Coarse Cut of Pillars by Using EDM

The lathe method (chapter (3.2.2)) includes two subsequent steps, the coarse cut of the pillar and the stepwise fine cut of the pillar. For large pillars with diameters larger than 20µm, milling the coarse shape of the pillar can take several hours, if it is done with the FIB. With electrical discharge machining (EDM) it is possible to machine the coarse cut much faster and in larger numbers.
The coarse cut of the sample in figure (3.5) was machined by using Wire EDM (also known as Spark EDM). A thin single-strand metall wire in conjunction with de-ionized water allows the wire to remove material from the workpiece. This is done by applying an electric field between the wire and the workpiece. As a spark jumps across the gap between the workpiece and the wire, material is removed from both electrodes. Since the workpiece is needed as an electrode, the material of the workpiece needs to be electrically conductive.

A.2 Indentation Strain Rate

In chapter (5.3.2) the indentation strain rate dependence of Fe is shown. The definition of the indentation strain rate is not as easy as in a tensile test, where the true strain rate can be determined according to:

$$\dot{\epsilon} = \frac{\dot{l}}{l} \tag{A.1}$$

$\dot{l}$ is the change of length and l denotes the actual length of the specimen.

For an indentation test the issue is more complex. In the following it is shown how the indentation strain rate can be determined, if a Berkovich indenter (projected area $A = 24.56 \cdot h^2$) is used.

The nanoindenter is inherently load-controlled, therefore a constant strain rate $\dot{\epsilon_{ind}}$ means that the ratio of load rate $\dot{P}$ and actual load P is kept constant. Assuming that the hardness H is constant over the entire indentation depth, the strain rate can be expressed by the ratio of indentation velocity $\dot{h}$ and actual depth h [111].

$$
\begin{aligned}
\dot{\epsilon}_{ind} = \frac{\dot{P}}{P} = \text{const.} \\
H = \text{const.} = \frac{P}{A} = \frac{P}{24.56 \cdot h^2} \\
\rightarrow P = 24.56 \cdot H \cdot h^2 \\
\dot{P}(h) = \frac{dP}{dh} \cdot \frac{dh}{dt} = 24.56 \cdot H \cdot 2h \frac{dh}{dt} \\
\frac{\dot{P}}{P} = \frac{24.56 \cdot H \cdot 2h \cdot \dot{h}}{24.56 \cdot H \cdot h^2} \\
\frac{\dot{P}}{P} = 2\frac{\dot{h}}{h}
\end{aligned}
\tag{A.2}
$$

The indentation strain rate of a Berkovisch indenter is therefore proportional to the ratio of indentation velocity $\dot{h}$ and indentation depth h.

A.3 Temperature Dependence of the Mechanical Size Effect of BCC Metals

From figures (6.3) and (6.2) can be seen that the mechanical size effect of bcc metals does not only depend on size, it also depends on the ratio of T_{test} and T_{ath}. The measurement at temperatures other than room temperature is challenging, since higher temperatures lead to thermal drift of the indenter tip and lower temperatures will result in the formation of ice on the sample. The temperature dependence of the power law exponents is linear as can be seen in figure (6.3) and from equation (6.1). This relationship and the measurements of the four bcc metals at room temperature can be used in order to estimate the flow stresses of these metals at different temperatures. The dependence on size and temperature is shown in figure (A.1). The black lines represent the size dependent behaviour of the four metals at room temperature (figure (6.2)).

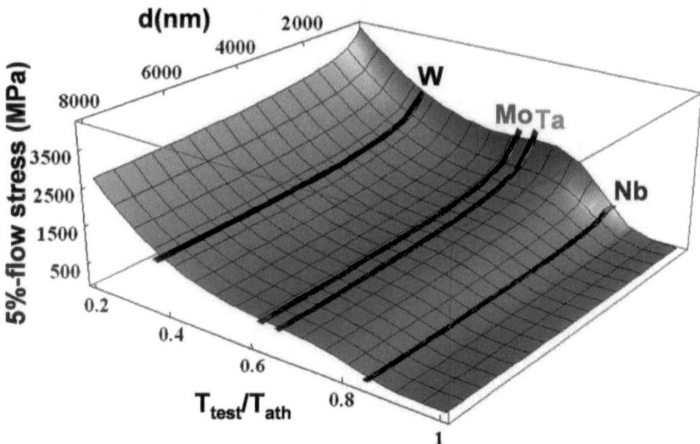

Figure A.1: Flow stress of W, Mo, Ta and Nb with respect to pillar diameter and $\frac{T_{test}}{T_{ath}}$.

A.4 Normalization by Shear Modulus Applied to BCC Metals

The mechanical size effects of Nb, Ta, Mb and W were compared and it was found that the size dependent behaviour depends on temperature (chapter (6.4)). Different normalization procedures have been applied in order to investigate this temperature dependence. FCC metals can be normalized by dividing the data by the shear modulus μ, as can be seen in figure (2.11). This normalization procedure was also applied to the four bcc metals, which is shown in figure (A.2). The flow stresses are just shifted to different values but the different slopes persist. This indicates that the temperature dependence of the mechanical size effect does not have its origin in the temperature dependence of the shear modulus.

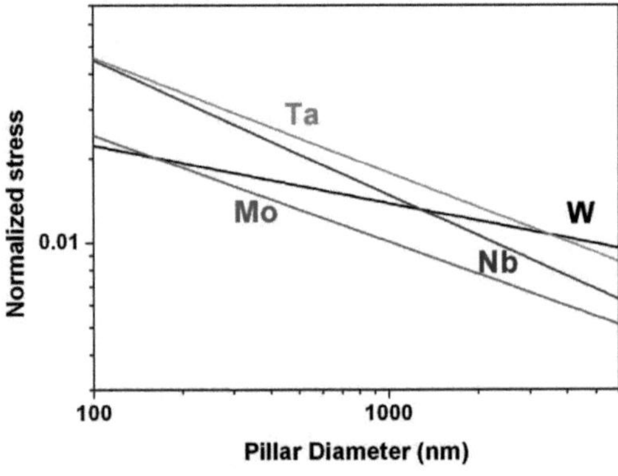

Figure A.2: Normalized flow stresses at 5% plastic of W, Mo, Ta and Nb with respect to pillar diameter.

A.5 Microcompression of Cuboidal Pillars: 'True Stress' Calculation

The microcompression of rectangular Ta pillars shows a difference in flow stress for the LS and SS configuration. This can either be a real effect of the material or it can be due to geometrical effects that cause higher true stresses in the SS configuration than in the LS configuration (chapter (6.6)). This issue has been reviewed by assuming a worst case scenario. It is assumed that the pillar does not show any elastic behaviour and the displacement of the indenter is completely transferred into the displacement of the pillar. Furthermore it was assumed that only one glide plane is activated and that the whole displacement occurs on only this single plane. The glide plane was assumed to have an inclination angle of 40° with respect to the sample surface. Figure (A.3) shows the results of this 'true'-stress analysis.

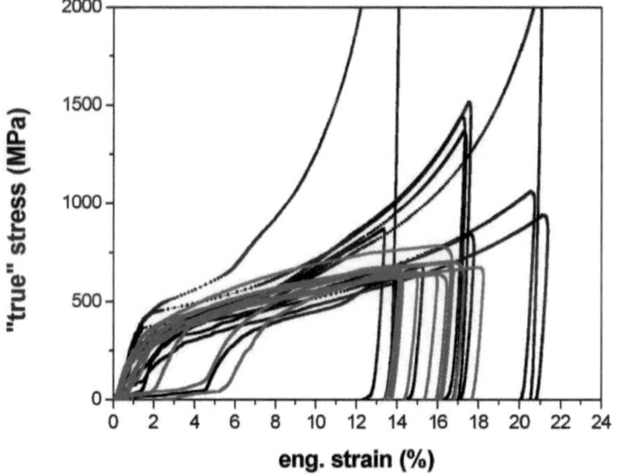

Figure A.3: Stress strain curves obtained from rectangular Ta pillars considering the different true stress for both orientations.

A comparison of the 5% flow stresses shows that even in this scenario the difference between parallel (LS) and perpendicular (SS) configuration still persists (figure (A.4)). The glide plane separates the pillar into an upper part and a lower part. If the indenter moves 500nm vertically down this results in a displacement of 778nm of the upper part with respect to the lower part. Such large displacements have never been observed and, therefore, the effect of the true stresses is certainly much weaker than shown in figure (A.3).

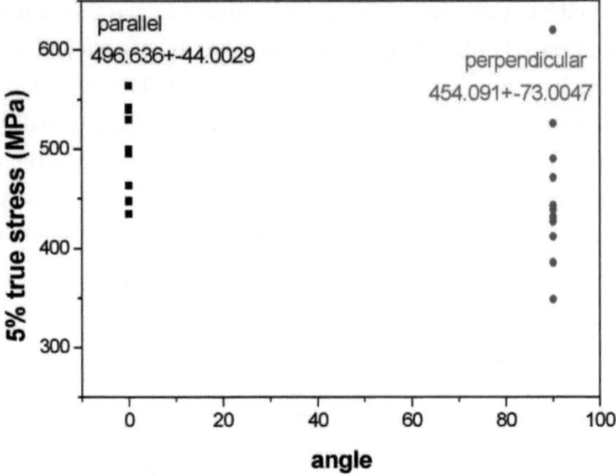

Figure A.4: 5% flow stresses for parallel (LS) and perpendicular (SS) configuration based on figure (A.3). On the x-axis the angle between the Burgers vector and the long edge of the cuboidal pillar is given.

Bibliography

[1] G.E. Dieter. Mechanical metallurgy. *McGraw-Hill, Columbus ,OH, USA*, page 124, 1961.

[2] L.P. Kubin, B. Devincre, and M. Tang. Mesoscopic modelling and simulation of plasticity in fcc and bcc crystals: dislocation intersections and mobility. *Journal of Computer-Aided Materials Design*, 5(1):31–54, 1998.

[3] U. Messerschmidt and M. Bartsch. Generation of dislocations during plastic deformation. *Materials Chemistry and Physics*, 81(2-3):518 – 523, 2003.

[4] W. Cai, V.V. Bulatov, S. Yip, and A.S. Argon. Kinetic Monte Carlo modeling of dislocation motion in bcc metals. *Materials Science and Engineering A*, 309-310:270 – 273, 2001.

[5] C.R. Weinberger and W. Cai. Surface-controlled dislocation multiplication in metal micropillars. *Proceedings of the National Academy of Sciences*, 105(38):14304–14307, 2008.

[6] M.D. Uchic, D.M. Dimiduk, J.N. Florando, and W.D. Nix. Sample dimensions influence strength and crystal plasticity. *Science*, 305(5686):986–989, 2004.

[7] M.D. Uchic, P.A. Shade, and D.M. Dimiduk. Plasticity of micrometer-scale single crystals in compression. *Annual Review of Materials Research*, 39(1):361–386, 2009.

[8] C.A. Volkert. unpublished results.

[9] C.A. Volkert and E.T. Lilleodden. Size effects in the deformation of submicron Au columns. *Philosophical Magazine*, 86(33-35):5567–5579, 2006.

[10] S. Brinckmann, J.Y. Kim, and J.R. Greer. Fundamental differences in mechanical behavior between two types of crystals at the nanoscale. *Physical Review Letters*, 100(15):155502, 2008.

[11] J.Y. Kim, D. Jang, and J.R. Greer. Insight into the deformation behavior of niobium single crystals under uniaxial compression and tension at the nanoscale. *Scripta Materialia*, 61(3):300 – 303, 2009.

[12] H. Matsui and H. Kimura. A mechanism of the unexpected {110} slip observed in bcc metals deformed at low temperatures. *Scripta Metallurgica*, 7:905–914, 1973.

[13] K.G. Hoge and A.K. Mukherjee. The temperature and strain rate dependence of the flow stress of tantalum. *Journal of Materials Science*, 12(8):1666–1672, 1977.

[14] C.N. Reid, A. Gilbert, and G.T. Hahn. Twinning, slip and catastrophic flow in niobium. *Acta Metallurgica*, 14(8):975 – 983, 1966.

[15] J.P. Hirth and J. Lothe. Theory of dislocations. *McGraw-Hill, Columbus, OH, USA*, pages 257–259, 1968.

[16] R. Maddin and N.K. Chen. Geometrical aspects of the plastic deformation of metal single crystals. *Progress in Metal Physics*, 5:53 – 95, 1954.

[17] T. Yalcinkaya, W.A.M. Brekelmans, and M.G.D. Geers. Bcc single crystal plasticity modeling and its experimental identification. *Modelling and Simulation in Materials Science and Engineering*, 16(8):085007 (16pp), 2008.

[18] E.N.C. Andrade and Y.S. Chow. The glide elements of body-centred cubic crystals, with special reference to the effect of temperature. *Proceedings of the Royal Society*, 175A:290, 1940.

[19] V. Vitek, M. Mrovec, R. Groeger, J. L. Bassani, V. Racherla, and L. Yin. Effects of non-glide stresses on the plastic flow of single and polycrystals of molybdenum. *Materials Science and Engineering A*, 387-389:138 – 142, 2004.

[20] M. S. Duesbery and V. Vitek. Plastic anisotropy in b.c.c. transition metals. *Acta Materialia*, 46(5):1481 – 1492, 1998.

[21] K. Ito and V. Vitek. Atomistic study of non-Schmid effects in the plastic yielding of bcc metals. *Philosophical Magazine A*, 81(5):1387–1407, 2001.

[22] R. Groeger and V. Vitek. Breakdown of the Schmid law in bcc molybdenum related to the effect of shear stress perpendicular to the slip direction. *Materials Science Forum*, 2005.

[23] S.N. Kuchnicki, R.A. Radovitzky, and A.M. Cuitio. An explicit formulation for multiscale modeling of bcc metals. *International Journal of Plasticity*, 24(12):2173 – 2191, 2008.

[24] A. Seeger and W. Wasserbaech. Anomalous slip - a feature of high-purity body-centred cubic metals. *physica status solidi (a)*, 189(1):27 – 50, 2002.

[25] J. Chaussidon, M. Fivel, and D. Rodney. The glide of screw dislocations in bcc Fe: Atomistic static and dynamic simulations. *Acta Materialia*, 54(13):3407 – 3416, 2006.

[26] H. Conrad. Conference on the relation between structure and strength in metals and alloys, teddington. 1963.

[27] B. Escaig. L'activation thermique des deviations sous faibles contraintes dans les structures h.c. et c.c par. *Phys. Stat. Solidi*, 28(463), 1968.

[28] M. Tang, L. P. Kubin, and G. R. Canova. Dislocation mobility and the mechanical response of b.c.c. single crystals: A mesoscopic approach. *Acta Materialia*, 46(9):3221 – 3235, 1998.

[29] J. S. Koehler. The nature of work-hardening. *Phys. Rev.*, 86(1):52–59, Apr 1952.

[30] E. Orowan. Dislocation in metals. *American Institute of Mining and Metallurgy Engineering, New York*, page 103, 1954.

[31] J.J.Gilman and W.G. Johnston. Behavior of individual dislocations in strain-hardened LiF crystals. *Journal of Applied Physics*, 31(4):687–692, 1960.

[32] M. Rhee, H.M. Zbib, J.P. Hirth, H. Huang, and T. de la Rubia. Models for long-/short-range interactions and cross slip in 3d dislocation simulation of bcc single crystals. *Modelling and Simulation in Materials Science and Engineering*, 6(4):467–492, 1998.

[33] J.P. Hirth and J. Lothe. Theory of dislocations. *McGraw-Hill, Columbus, OH, USA*, pages 67–68, 1968.

[34] J.R. Greer, C.R. Weinberger, and W. Cai. Comparing the strength of f.c.c. and b.c.c. sub-micrometer pillars: Compression experiments and dislocation dynamics simulations. *Materials Science and Engineering: A*, 493(1-2):21 – 25, 2008.

[35] E. Arzt. Size effects in materials due to microstructural and dimensional constraints: a comparative review. *Acta Materialia*, 46(16):5611 – 5626, 1998.

[36] D.M. Dimiduk, M.D. Uchic, S.I. Rao, C. Woodward, and T.A. Parthasarathy. Overview of experiments on microcrystal plasticity in fcc-derivative materials: selected challenges for modelling and simulation of plasticity. *Modelling and Simulation in Materials Science and Engineering*, 15(2):135, 2007.

[37] W.D. Nix. Mechanical properties of thin films. *Metallurgical and Materials Transactions A*, 20(11):2217 – 2245, 1988.

[38] O. Kraft, P.A. Gruber, R. Moenig, and D. Weygand. Plasticity in confined dimensions. *Annual Review of Materials Research*, 40(1):293–317, 2010.

[39] J.P. Hirth and J. Lothe. Theory of dislocations. *McGraw-Hill, Columbus, OH, USA*, pages 716–718, 1968.

[40] H.W. Song, S.R. Guo, and Z.Q. Hu. A coherent polycrystal model for the inverse hall-petch relation in nanocrystalline materials. *Nanostructured Materials*, 11(2):203 – 210, 1999.

[41] R. Venkatraman and J.C. Bravman. Separation of film thickness and grain boundary strengthening effects in Al thin films on Si. *Journal of Materials Research*, 7(8):2040–2048, 1992.

[42] W.D. Nix. Yielding and strain hardening of thin metal films on substrates. *Scripta Materialia*, 39(4-5):545 – 554, 1998.

[43] O. Kraft, M. Hommel, and E. Arzt. X-ray diffraction as a tool to study the mechanical behaviour of thin films. *Materials Science and Engineering A*, 288(2):209 – 216, 2000.

[44] C.V. Thompson. The yield stress of polycrystalline thin films. *Journal of Materials Research*, 8(2):237–238, 1993.

[45] R.M. Keller, S.P. Baker, and E. Arzt. Quantitative analysis of strengthening mechanisms in thin cu films: Effects of film thickness, grain size and passivation. *Journal of Materials Research*, 13(5):1307–1317, 1998.

[46] H. Bei, S. Shim, E.P. George, M.K. Miller, E.G. Herbert, and G.M. Pharr. Compressive strengths of molybdenum alloy micro-pillars prepared using a new technique. *Scripta Materialia*, 57(5):397 – 400, 2007.

[47] S. Buzzi, M. Dietiker, K. Kunze, R. Spolenak, and J.F. Loeffler. Deformation behaviour of silver submicrometer-pillars prepared by nanoimprinting. *Philosophical Magazine*, 89(10):869 – 884, 2009.

[48] Z.W. Shan, R.K. Mishra, S.A. Syed Asif, O.L. Warren, and A.M. Minor. Mechanical annealing and source-limited deformation in submicrometre-diameter Ni crystals. *Nature Materials*, 7:115–119, 2007.

[49] D.M. Dimiduk, M.D. Uchic, and T.A. Parthasarathy. Size-affected single-slip behavior of pure nickel microcrystals. *Acta Materialia*, 53(15):4065 – 4077, 2005.

[50] C.P. Frick, B.G. Clark, S. Orso, A.S. Schneider, and E. Arzt. Size effect on strength and strain hardening of small-scale [111] nickel compression pillars. *Materials Science and Engineering: A*, 489(1-2):319 – 329, 2008.

[51] J.R. Greer and W.D. Nix. Nanoscale gold pillars strengthened through dislocation starvation. *Physical Review B (Condensed Matter and Materials Physics)*, 73(24):245410, 2006.

[52] J.R. Greer, W.C. Oliver, and W.D. Nix. Size dependence of mechanical properties of gold at the micron scale in the absence of strain gradients. *Acta Materialia*, 53(6):1821 – 1830, 2005.

[53] J.R. Greer and W.D. Nix. Size dependence of mechanical properties of gold at the sub-micron scale. *Applied Physics A: Materials Science and Processing*, 80(8):1625 – 1629, 2005.

[54] K.S. Ng and A.H.W. Ngan. Stochastic nature of plasticity of aluminum micropillars. *Acta Materialia*, 56(8):1712 – 1720, 2008.

[55] D. Kiener. Size effects in single crystal plasticity of copper under uniaxial loading. *Dissertation, Montanuniversitaet Leoben, AT*, 2007.

[56] D. Kiener, C. Motz, T. Schoberl, M. Jenko, and G. Dehm. Determination of mechanical properties of copper at the micron scale. *Advanced Engineering Materials*, 8, 2006.

[57] D.M. Norfleet, D.M. Dimiduk, S.J. Polasik, M.D. Uchic, and M.J. Mills. Dislocation structures and their relationship to strength in deformed nickel microcrystals. *Acta Materialia*, 56(13):2988 – 3001, 2008.

[58] J.P. Hirth and J. Lothe. Theory of dislocations. *McGraw-Hill, Columbus, OH, USA*, pages 6–7, 1968.

[59] H. Bei, S. Shim, G.M. Pharr, and E.P. George. Effects of pre-strain on the compressive stress-strain response of mo-alloy single-crystal micropillars. *Acta Materialia*, 56(17):4762 – 4770, 2008.

[60] S.I. Rao. Estimating the strength of single-ended dislocation sources in micron-sized single crystals. *Philosophical Magazine*, 87:4777–4794(18), October 2007.

[61] A. Needleman L. Nicola, E. Van Der Giessen. Size effects in polycrystalline thin films analyzed by discrete dislocation plasticity. *Thin Solid Films*, 479(1-2):329–338, 2005.

[62] A. Needleman L. Nicola, E. Van der Giessen. Discrete dislocation analysis of size effects in thin films. *Journal of Applied Physics*, 93(10 1):5920–5928, 2003.

[63] V.S. Deshpande, A. Needleman, and E. Van der Giessen. Plasticity size effects in tension and compression of single crystals. *Journal of the Mechanics and Physics of Solids*, 53(12):2661 – 2691, 2005.

[64] J.J. Vlassak E. Van der Giessen A. Needleman L. Nicola, Y. Xiang. Plastic deformation of freestanding thin films: Experiments and modeling. *Journal of the Mechanics and Physics of Solids*, 54(10):2089–2110, 2006.

[65] W. Cai C.R. Weinberger. Computing image stress in an elastic cylinder. *Journal of the Mechanics and Physics of Solids*, 55(10):2027–2054, 2007.

[66] S. Soleymani Shishvan, S. Mohammadi, and M. Rahimian. A dislocation-dynamics-based derivation of the frankread source characteristics for discrete dislocation plasticity. *Modelling and Simulation in Materials Science and Engineering*, 16(7):075002, 2008.

[67] E. Van Der Giessen A. Needleman D. Weygand, L.H. Friedman. Aspects of boundary-value problem solutions with three-dimensional dislocation dynamics. *Modelling and Simulation in Materials Science and Engineering*, 10(4):437–468, 2002. cited By (since 1996) 55.

[68] P. Gumbsch D. Weygand. Study of dislocation reactions and rearrangements under different loading conditions. *Materials Science and Engineering A*, 400-401(1-2 SUPPL.):158–161, 2005. cited By (since 1996) 10.

[69] J. Senger, D. Weygand, P. Gumbsch, and O. Kraft. Discrete dislocation simulations of the plasticity of micro-pillars under uniaxial loading. *Scripta Materialia*, 58(7):587 – 590, 2008.

[70] S.I. Rao, D.M. Dimiduk, T.A. Parthasarathy, M.D. Uchic, M. Tang, and C. Woodward. Athermal mechanisms of size-dependent crystal flow gleaned from three-dimensional discrete dislocation simulations. *Acta Materialia*, 56(13):3245 – 3259, 2008.

[71] T.A. Parthasarathy, S.I. Rao, D.M. Dimiduk, M.D. Uchic, and D.R. Trinkle. Contribution to size effect of yield strength from the stochastics of dislocation source lengths in finite samples. *Scripta Materialia*, 56(4):313 – 316, 2007.

[72] J.P. Hirth and J. Lothe. Theory of dislocations. *McGraw-Hill, Columbus, OH, USA*, pages 129–130, 1968.

[73] S. Shim, H. Bei, M.K. Miller, G.M. Pharr, and E.P. George. Effects of focused ion beam milling on the compressive behavior of directionally solidified micropillars and the nanoindentation response of an electropolished surface. *Acta Materialia*, 57(2):503 – 510, 2009.

[74] H. Bei, S. Shim, M.K. Miller, G.M. Pharr, and E.P. George. Effects of focused ion beam milling on the nanomechanical behavior of a molybdenum-alloy single crystal. *Applied Physics Letters*, 91(11):111915, 2007.

[75] C.M. Byer, B. Li, B. Cao, and K.T. Ramesh. Microcompression of single-crystal magnesium. *Scripta Materialia*, 62(8):536 – 539, 2010.

[76] T. Kitahara, S. Ando, M. Tsushida, H. Kitahara, and H. Tonda. Deformation behaviour of magnesium single crystals in c-axis compresion. *Key Engineering Materials*, 345-346:129 – 132, 2007.

[77] T. Obara, H. Yoshinga, and S. Morozumi. $\{1122\}\langle\bar{1}\bar{1}23\rangle$ slip system in magnesium. *Acta Metallurgica*, 21(7):845 – 853, 1973.

[78] Q. Yu, Z.-W. Shan, J. Li, X. Huang, L. Xiao, J. Sun, and E. Ma. Strong crystal size effect on deformation twinning. *Nature*, 463:335–338, 2009.

[79] A. S. Schneider, D. Kaufmann, B. G. Clark, C. P. Frick, P. A. Gruber, R. Moenig, O. Kraft, and E. Arzt. Correlation between critical temperature and strength of small-scale bcc pillars. *Physical Review Letters*, 103(10):105501, 2009.

[80] A.S. Schneider, B.G. Clark, C.P. Frick, P.A. Gruber, and E. Arzt. Effect of orientation and loading rate on compression behavior of small-scale mo pillars. *Materials Science and Engineering: A*, 508(1-2):241 – 246, 2009.

[81] N. M. Jennett, R. Ghisleni, and J. Michler. Enhanced yield strength of materials: The thinness effect. *Applied Physics Letters*, 95(12):123102, 2009.

[82] W.C. Oliver and G.M. Pharr. An improved technique for determining hardness and elastic modulus using load and displacement sensing indentation experiments. *Journal of Materials Research*, 7(6):1564–1583, 1992.

[83] D. Vesely. The influence of the surface orientation on yield in Mo single crystals. *Scripta Metallurgica*, 6(8):753 – 755, 1972.

[84] J.J. Vlassak and W.D. Nix. Measuring the elastic properties of anisotropic materials by means of indentation experiments. *Journal of the Mechanics and Physics of Solids*, 42(8):1223–1245, 1994.

[85] W. A. Brantley. Calculated elastic constants for stress problems associated with semiconductor devices. *Journal of Applied Physics*, 44(1):534–535, 1973.

[86] M. Werner. Temperature and strain-rate dependence of the flow stress of ultrapure tantalum single crystals. *physica status solidi (a)*, 104(1):63–78, 1987.

[87] G. Richter, K. Hillerich, D.S. Gianola, R. Moenig, O. Kraft, and C.A. Volkert. Ultrahigh strength single crystalline nanowhiskers grown by physical vapor deposition. *Nano Letters*, 9(8):3048–3052, 2009.

[88] S.A. Syed Asif and J.B. Pethica. Nanoindentation creep of single-crystal tungsten and gallium arsenide. *Philosophical magazine*, 76(6):1105–1118, 1997.

[89] A.C. Fischer-Cripps. Nanoindentation. *Mechanical Engineering Series, Springer New York*, page 41, 2002.

[90] G.E. Fougere, L. Riester, M. Ferber, J.R. Weertmann, and R.W. Siegel. Young's modulus of nanocrystalline Fe measured by nanoindentation. *Materials Science and Engineering A*, 204:1–6, 1995.

[91] J.P. Hirth and J. Lothe. Theory of dislocations. *McGraw-Hill, Columbus, OH, USA*, pages 682–685, 1968.

[92] H.G. Frost and M.F. Ashby. Deformation mechanism maps, chapter 18. *Pergamon Press*, 1982.

[93] C. Domain and G. Monnet. Simulation of screw dislocation motion in iron by molecular dynamics simulations. *Phys. Rev. Lett.*, 95(21):215506, 2005.

[94] W.A. Spitzig and A.S. Keh. The effect of orientation and temperature on the plastic flow properties of iron single crystals. *Acta Metallurgica*, 18(6):611–622, 1970.

[95] E. Kuramoto, Y. Aono, and K. Kitajima. Thermally activated slip deformation of high purity iron single crystals between 4.2k and 300k. *Scripta Metallurgica*, 13(11):1039–1042, 1979.

[96] D.F. Stein, J.R. Low Jr., and A.U. Seybolt. The mechanical properties of iron single crystals containing less than $5 \cdot 10^{-3}$ppm carbon. *Acta Metallurgica*, 11(11):1253–1262, 1963.

[97] A. Seeger. Slip planes and kink properties of screw dislocations in high-purity niobium. *Philosophical Magazine*, 86:3861–3892(32), -26/1-11 September 2006.

[98] L. Hollang, M. Hommel, and A. Seeger. The flow stress of ultra-high-purity molybdenum single crystals. *physica status solidi (a)*, 160(2):329–354, 1997.

[99] D. Brunner and V. Glebovsky. Analysis of flow-stress measurements of high-purity tungsten single crystals. *Materials Letters*, 44(3-4):144 – 152, 2000.

[100] R. Groeger and V. Vitek. Explanation of the discrepancy between the measured and atomistically calculated yield stresses in body-centred cubic metals. *Philosophical Magazine Letters*, 87(2):113–120, 2007.

[101] D. Kaufmann, A.S. Schneider, R. Moenig, C.A. Volkert, and O. Kraft. Effect of surface orientation on the plasticity of small bcc metalls. *to be submitted to Acta Materialia.*

[102] D. Vesely. The study of slip bands on the surface of Mo single crystals. *Physica Status Solidi*, 29:685–696, 1968.

[103] H. Matsui, H. Kimura, H. Saka, K. Noda, and T. Imura. Anomalous slip induced by the surface effect in molybdenum single-crystal foils deformed in a high voltage electron microscope. *Materials Science and Engineering*, 53:263–272, 1982.

[104] R.A. Masumura, P.M. Hazzledine, and C.S. Pande. Yield stress of fine grained materials. *Acta Materialia*, 46(13):4527 – 4534, 1998.

[105] Q. Wei, S. Cheng, K. T. Ramesh, and E. Ma. Effect of nanocrystalline and ultrafine grain sizes on the strain rate sensitivity and activation volume: fcc versus bcc metals. *Materials Science and Engineering A*, 381(1-2):71 – 79, 2004.

[106] R.J. Asaro and S. Suresh. Mechanistic models for the activation volume and rate sensitivity in metals with nanocrystalline grains and nano-scale twins. *Acta Materialia*, 53(12):3369 – 3382, 2005.

[107] G. Taylor. Thermally activated deformation of bcc metals and alloys. *Progress in Materials Science*, 36:29–61, 1992.

[108] J.P. Hirth and J. Lothe. Theory of dislocations. *McGraw-Hill, Columbus, OH, USA*, pages 538–540, 1968.

[109] D. Ostwaldt, J.R. Klepaczko, and P. Klimanek. Compression tests of polycrystalline α-iron up to high strains over a large range of strain rates. *Journal de Physique IV*, 7, 1997.

[110] M.G. da Silva and K.T. Ramesh. The rate-dependent deformation of porous pure iron. *International Journal of Plasticity*, 13(6-7):587–610, 1997.

[111] G.M. Pharr. private communication. 2010.